AN ILLUSTRATED HISTORY OF DUSTCARTS

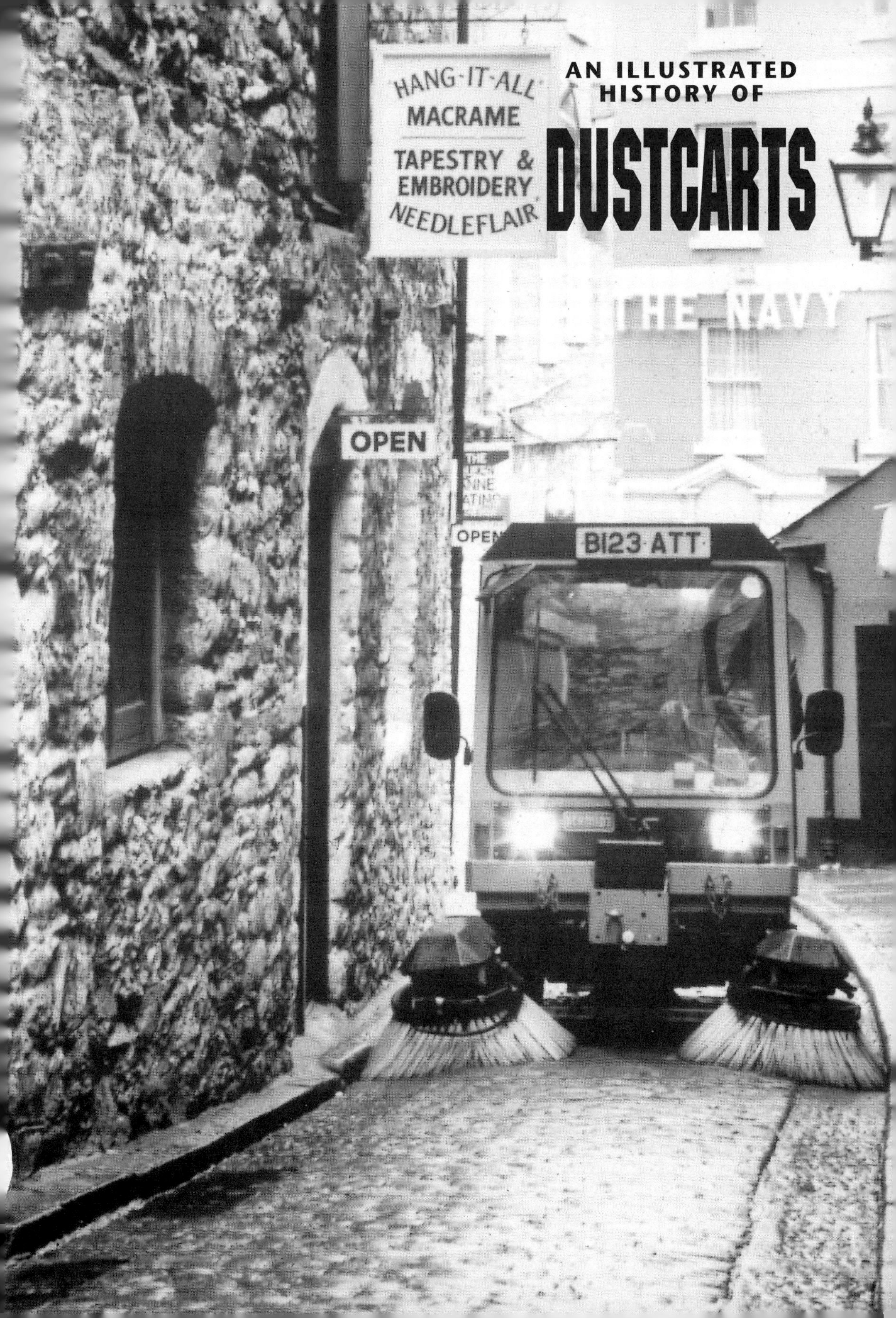

LORRY
RLW 945

AN ILLUSTRATED HISTORY OF DUSTCARTS

HINTON J. SHERYN

Front cover: Operated here by the Borough of Reigate and Banstead, the 2-11 model was an updated version of the Seddon Atkinson Series. It was first produced in 1991 and powered by a Perkins Phaser engine of 160hp. It had a new cab styling and is fitted here with a side-loading automatic collection system.
Seddon Atkinson archives

Back cover (top): This redesigned 75-gallon Karrier Yorkshire gully emptier is mounted on the short wheelbase (9ft 3in) CK3 chassis in April 1946. *Ian Allan Library*

Back cover (bottom): A Shelvoke & Drewry 'Freighter' refuse collection vehicle at work in Manchester on 1 April 1938. *Mike Pilkington, Manchester City Council*

Half title: A Schmidt street cleaner dating from the early 1980s.
Johnston Engineering Ltd

Title page: The Commer Karrier Bantam, seen here in September 1957, is being operated by the Borough of Wembley to collect and empty street sweepings from bins carried by pedestrian-operated electric orderlies. A Nel lorry loader is being used to hoist the bins. *Ian Allan Library*

First published 2000

ISBN 0 7110 2755 2

Published by Ian Allan Publishing

an imprint of Ian Allan Publishing Ltd, Terminal House, Shepperton, Surrey TW17 8AS.
Printed by Ian Allan Printing Ltd, Riverdene Business Park, Hersham, Surrey KT12 4RG.

Code: 0011/B

CONTENTS

PICTURE CREDITS

The author would like to thank the following organisations and individuals for their help in compiling the illustrations for this book:

Dennis Eagle (R. D. Taylor), Heathcote Way, Heathcote Ind Est, Warwick CV34 6TE;

ERF Ltd, Sun Works, Sandbach, Cheshire;

Faun Municipal Vehicles Ltd, Llangefni, Anglesey, North Wales;

Mr Colin Humble, Torquay;

Ian Allan Publishing Archives;

Institute of Waste Management;

Johnston Engineering Ltd, Curtis Road, Dorking, Surrey;

LWS Services (Tony Brown), Lancashire House, 24 Winckley Square, Preston PR1 3JJ;

J. McArdle Contracts Ltd (Mark Harvey), Wembley;

Manchester City Council;

Seddon Atkinson Ltd (Richard Grey), Woodstock Factory, Oldham, Lancs;

G. & J. Seddon Ltd, Armitage Avenue, Little Hulton, Worsley, Manchester M38 OFH;

Mr Bernard Titcombe, Staines, Middlesex

THE EARLY YEARS

The history of dustcarts and street cleaning can be traced back several centuries. By the time of the 17th century in Britain, it was becoming increasingly difficult to cleanse the city streets and dispose of the hundreds of tons of waste created by every household in the land. Following the Great Fire of London in 1666, the authorities realised how much of a hazard the accumulation of rubbish had become and began to take a keener interest in street cleaning and waste disposal but despite talks held between the Alderman and civic officials, the problem in the 1700s was still about as bad as it could be.

The almost universal use of horses for the transportation of both goods and people also created its own problems, with ton upon ton of horse manure being dumped upon the streets on a daily basis. Coupled with horses urinating everywhere and the slop pails being emptied into the streets from almost every household, the recipe for filth, slime, rubbish and disease amongst the population was already laid down, especially when added to the malnutrition and lack of the basic amenities we all now take for granted.

Although during the 18th century road builders had begun experimenting with the use of tarmacadam — using a tar coating on road surfaces — most roads were surfaced with cobbled stone, which made cleaning very difficult. Many towns had also been using a coating of stone sets made from granite or slate, sometimes referred to as York slabs, and granite and other stones were used to produce chippings for certain areas of pathway but, again, cleaning such surfaces was extremely difficult. However, the councils who were concerned with the general welfare of the inhabitants of their boroughs did set about employing cleaners of city streets and pathways — such as they were in those far-off days.

Only in the 19th century was the task of cleansing the streets really taken seriously. Indeed, one Health Officer for London in the 19th century, G. F. Willoughby MD, devised a pocket-book which would help identify the source of accumulated dirt and dust on the streets of London and Cambridge, and how best it could be disposed of: 'The refuse from streets both in dry and wet weather consists of the debris of the paving itself, iron from horse-shoes, powdered horse dung in varying proportions. The moisture content varies from 35% to 90%. The dry solids average from 55% to 60% of horse dung, 30% to 35% of powdered stone (paving material) and 10% to 15% of abraded iron.'

A later analysis, made in 1867 by Dr Letherby, of mud from London's stone-paved streets and of horse droppings and farmyard manure (from animals brought into the town's markets for sale) dried at a temperature of from 266 to 300° Fahrenheit, is as follows:

			Mud From Paved Streets	
Constituents	Fresh Horse Droppings	Farmyard Dung	Max Organic Dry Weather	Min Organic Wet Weather
	%	%	%	%
Organic	82.7	69.9	58.2	20.5
Mineral	17.3	30.1	41.8	79.5
	100.0	100.0	100.0	100.0

Dried mud taken from wood pavements (which were fairly common) produced approximately 60% of organic matter. The disposal of much of the organic waste was less of a problem in that by placing it on farmland it proved to be of benefit to the farmers as well as to the community.

General household rubbish was normally found in streams, rivers and in lanes; also occasionally in 'laystalls' which were centres set up for collection. Indeed, Victorian dumps or rubbish tips are frequently unearthed today during the excavation of sites for house-building or infrastructure work. Where practical, much of the country's waste was disposed of into the seas around our coastline. Indeed, some of our major cities adopted the method of tipping their refuse directly into the sea from hopper barges. This could create a nuisance when the return tide threw the debris back on to the beaches if the disposal point at sea was not far enough out to avoid this happening.

In many cases, household rubbish consisting of vegetable matter was mixed in with street sweepings and disposed of to local farmers. This, however, necessitated the separation of tins, china, glass and other matter of no use to farmers for manurial purposes.

The rakers and cleansers of the 18th century, equipped with their primitive brushes, scoops and rakes, made little impact on the plight of many of our cities so in 1817 the Metropolitan Paving Act was passed for London, in an effort to enforce the regular cleansing of the streets. Special bodies of commissioners, or 'vestries', were instructed to allow contractors to remove refuse at their own expense. This was certainly no barrier, and many contractors lined up to participate, such were the profits available from refuse.

The first mechanical street sweeping equipment was also developed at that time, in the early years of the 19th century. In 1828 a horse-drawn machine with a water tank, scraper and a fixed brush of heather, was patented by Messrs Boase and Smith of London. The idea was later developed by William Smith, who added a rotating brush on his horse-drawn patented sweeper. The idea was to move rubbish and dirt into windrows (a row of piles set up for drying) at the side of the road, which could be collected by contractors using shovels and carts. Separately, in 1843, a mechanical sweeper was patented by Joseph Whitworth of Manchester. This consisted of a conveyor belt which picked up the dirt using a series of 'squeegees'. The belt was driven by a sprocket and chains connected to the wheels of the cart. The rubbish was aimed into the cart for disposal, when full, at an authorised dumping point. By the mid-19th century the horse and cart provided the usual means for collecting and disposing of household waste in built-up areas.

The same authorities who are responsible for street cleaning have also long had to clear the streets of snow during the winter. During inclement weather, many of the street cleaners would often find themselves clearing snow and ice from pedestrian and road traffic ways. Little has changed in this respect today.

Sand was widely used to provide added traction in icy weather conditions. Specially-built wagons and carts were employed to distribute the sand over the streets in an organised way, so as not to leave hazardous lumps of sand in place of the snow or ice.

Above: A drawing showing the street orderly shovels and brushes used on the mainly asphalt and wooden-paved roads of the time. *Bernard Titcombe collection*

Above:
A street orderly boy using a simple dust pan and handbrush to clean London Corporation's streets. Their use was widespread in the 18th century and up until the mid-19th century, before more efficient mechanical methods were adopted.
Bernard Titcombe collection

Above and below:
The first sweeping machine, as made by Smith & Sons, Barnard Castle. It was necessary to scrape the road surface to loosen the embedded matter and in winter to remove snow from the walkways and streets. This led to many types of horse-drawn scrapers (or graders) being developed, one of which is featured below. *Bernard Titcombe collection*

Left:
An essential part of street cleaning was that of washing the roads using water hoses to remove all traces of unpleasant matter such as urine (from horses, dogs and people). The street flushers were kitted out with a cart, brushes, schivers (a type of scraping blade mounted at the bottom end of a broom handle), a hose and a series of running trollies (over which the hose could run to save it from wear on the street surface, also enabling it to be controlled more easily). *Bernard Titcombe collection*

Left:
A variety of types of barrow were used by street orderlies. The one illustrated here was the type chosen by the Borough of Paddington in London, made by The Bristol Wagon & Carriage Works Co Ltd.
Bernard Titcombe collection

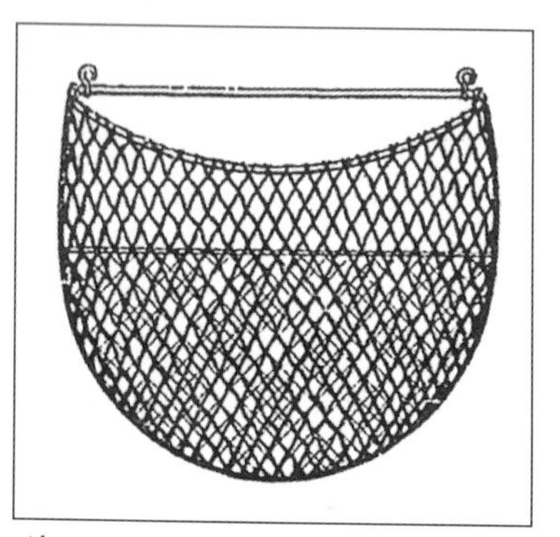

Above:
Galvanised iron wire baskets made for the reception of waste paper, orange peel, etc were fixed to railings and walls, to be emptied by the street cleaners during their rounds.
Bernard Titcombe collection

Right:
Bins were also sunk into the pavements or streets to ground level, so as not to become hazardous to pedestrians or traffic. They were emptied by the street orderly during his rounds, by lifting the cast-iron lid and removing a galvanised box containing the litter.
Bernard Titcombe collection

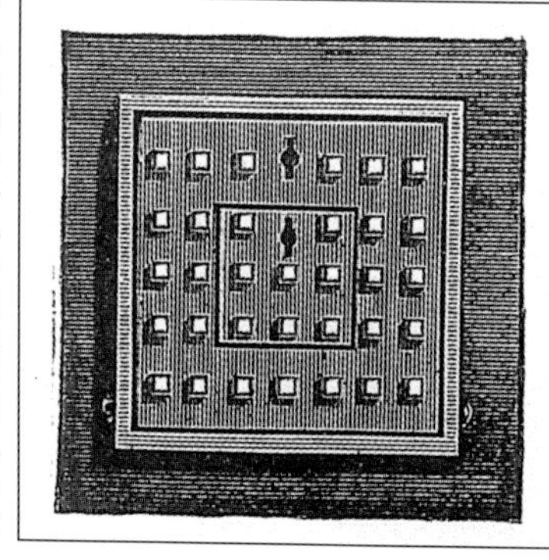

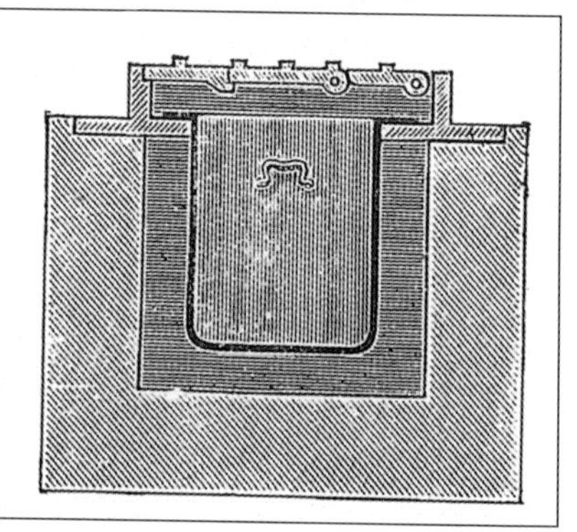

Right:
A dust van built by the Horsfall Company with a low loading side so that a ladder would not be required. *Bernard Titcombe collection*

Left:
A hand watering truck, used for watering courts and pavements. *Bernard Titcombe collection*

Below:
The horse and cart provided the means for collecting and disposing of household waste until the motorised carts took over. This photograph was taken on 23 September 1911 at the Water Street Cleaning Yard in Manchester.
Manchester City Council

Left: A horse-drawn street watering wagon, built by The Bristol Wagon & Carriage Works. *Bernard Titcombe collection*

Right: The Bristol Wagon & Carriage Works Ltd, of Lawrence Hill, Bristol, advertises its horse-drawn vehicles. *Bernard Titcombe collection*

Below: John Smith & Sons were builders of Smith's improved end-tipping vans and other horse-drawn carts and implements. *Bernard Titcombe collection*

Established 1794.

CONTRACTORS TO

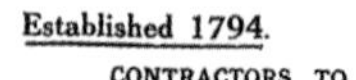

HIS MAJESTY'S GOVERNMENT.

Contractors to PRINCIPAL MUNICIPAL CORPORATIONS, and COUNTY COUNCILS.

ON ADMIRALTY LIST.

Telegrams: SMITH, INKERMANN STREET, WOLVERHAMPTON.

TELEPHONES { 700 Wolverhampton. 7191 Central B'Ham.

John Smith & Sons,

Inkermann Street,

Wolverhampton,

AND

Pritchett Street,

Birmingham.

BUILDERS of Watercarts, Tumbler Carts, Sweeping-Machines and Tipping Carts and Vans for Municipal Work and every class of Commercial Work.

Smith's improved End-tipping Vans.

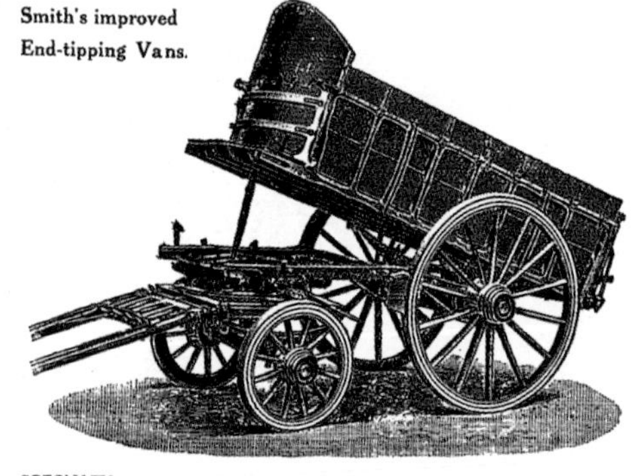

SPECIALTY:—

OUR PATENT SIDE TIPPING DUST OR SLOP VAN.

THE

BRISTOL WAGON AND CARRIAGE WORKS, LTD.

LAWRENCE HILL, BRISTOL.

Patent Screw Tip DUST WAGONS with our New Patent Automatic COVERS.

The Best DUST WAGON on the MARKET.

Best Material and Workmanship Guaranteed.

Awarded Prize Medal Royal Sanitary Institution, Health Exhibition, Bristol 1906.

Makers of:

WATER VANS & CARTS, IRON BODIED SANITARY TUMBLER CARTS,

And all Vehicles for Street Cleansing Purposes.

MOTOR AND HORSE AMBULANCE VANS,

DISINFECTING VANS AND CARTS.

(Chief Offices and Works,) **LAWRENCE HILL, BRISTOL.**

Showrooms: VICTORIA STREET, BRISTOL.

London Offices: 154, SUFFOLK HOUSE, LAURENCE POUNTNEY HILL, E.C.

London Repairing Dept.: MANTELL STREET (late Sermon Lane), ISLINGTON, N.

Telegrams: "WAGON," BRISTOL "BRISTOCAR," LONDON.

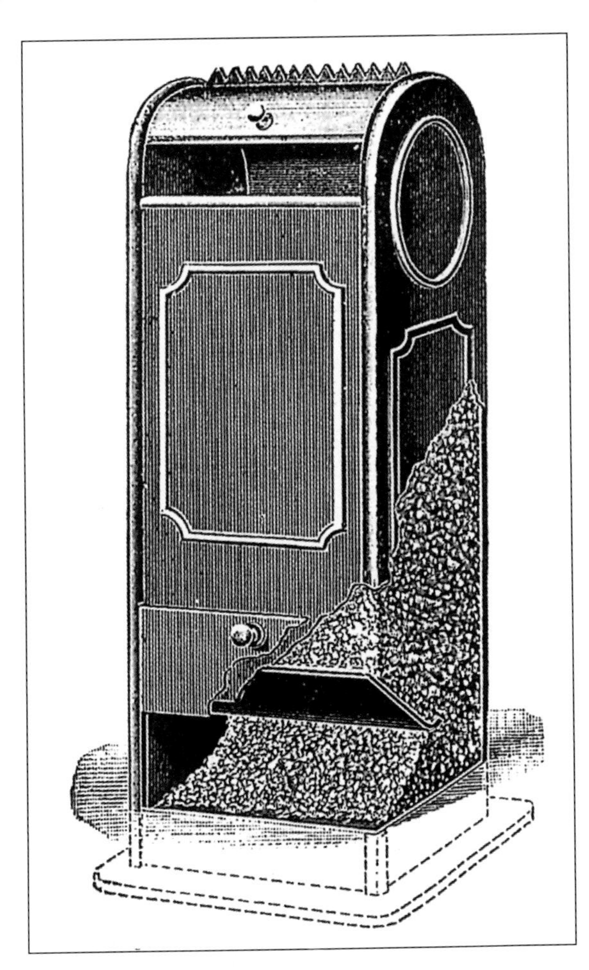

Left:
An improved sand bin, used to contain sand for broadcasting on icy roads and pavements. *Bernard Titcombe collection*

Above:
A galvanised iron sand bin, used for the same purpose. *Bernard Titcombe collection*

Left:
A mechanical gully emptier, basically a hand-operated pump. *Bernard Titcombe collection*

Below:
An open-type iron sand bin.
Bernard Titcombe collection

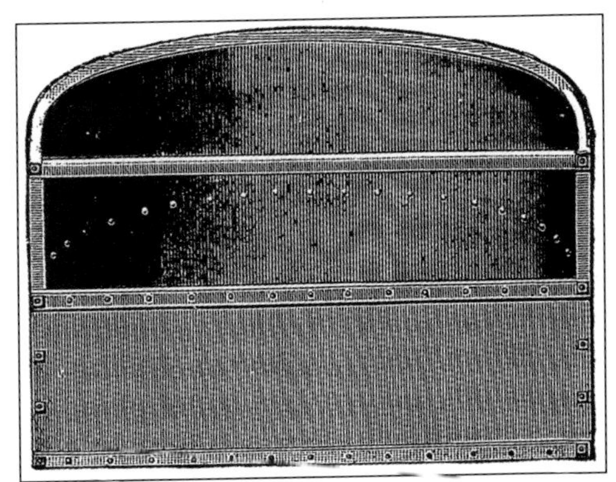

Left:
This sand distributor cart employed an Archimedean screw to empty it of its load of sand. *Bernard Titcombe collection*

Right:
A heavy snowfall such as this one which hit Aberdeen in 1898 would need clearing by the authorities. *Bernard Titcombe collection*

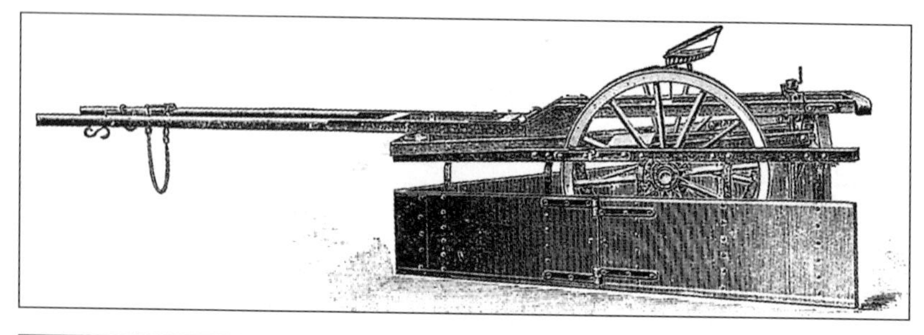

Left:
Example of a horse-drawn snow plough. *Bernard Titcombe collection*

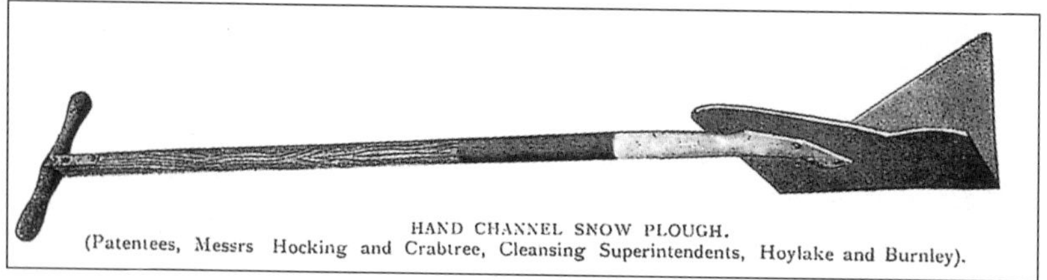

HAND CHANNEL SNOW PLOUGH.
(Patentees, Messrs Hocking and Crabtree, Cleansing Superintendents, Hoylake and Burnley).

Left:
Example of a hand channel snow plough. *Bernard Titcombe collection*

Incinerators and Liquid Waste

As the disposal of waste matter became somewhat more civilised and organised, slop carts were employed to take away the contents of the chamber pots and lavatory pails, much of it to specially built tanks strategically placed throughout London and the other big cities. These large tanks provided the only economical method of disposing of waste matters which contained between 40% and 90% water for every ton of slop. Although they were found to be expensive to construct initially, they had the advantage of being always ready for use, whereas cheaper alternatives such as the building of dams provided only temporary solutions.

The tanks were constructed either of wood or concrete and were generally built on or above the level of the ground, though on occasions they were sunk into the ground, depending on the method being considered as best for tipping the carts when they arrived. A major consideration was to ensure a rapid and complete drainage into the already constructed underground sewer systems.

Basically, the liquids were allowed to drain away into the underground system, while the solids remained to solidify. They were then ready to be removed by men using shovels or mechanical grabs, being loaded into carts which took the material to farmland or to local brick works (where it would be mixed with dust and baked in kilns).

The destruction of household waste by burning or incinerating has long been one of the most widely used methods, providing the hazardous particulates are disposed of into the atmosphere by the chimneys at such plants. When many of these plants were established in the early years of the 20th century, both in Britain and abroad, such as the Cairo plant which opened in 1904 and is illustrated in this chapter, few, if any, plastics were encountered which might have had dire consequences for local residents.

SLOP TUMBLER CART.

Examples of a slop tumbler cart, a slop wagon made by Horsfall and a plain open-plank-sided slop wagon made by The Bristol Wagon Company. What happened to the contents of these carts? Much was taken to specially built tanks strategically placed throughout London and the big cities. *Bernard Titcombe collection*

Above: Example of a built-up slop tank at Finsbury in North London. *Bernard Titcombe collection*

Above: A typical dust or slop cart manufactured by the Bristol Wagon Company. *Bernard Titcombe collection*

Above: An example of a horse-drawn night soil wagon. *Bernard Titcombe collection*

Left: This horse-drawn fish offal wagon was employed by Finsbury Borough Council. *Bernard Titcombe collection*

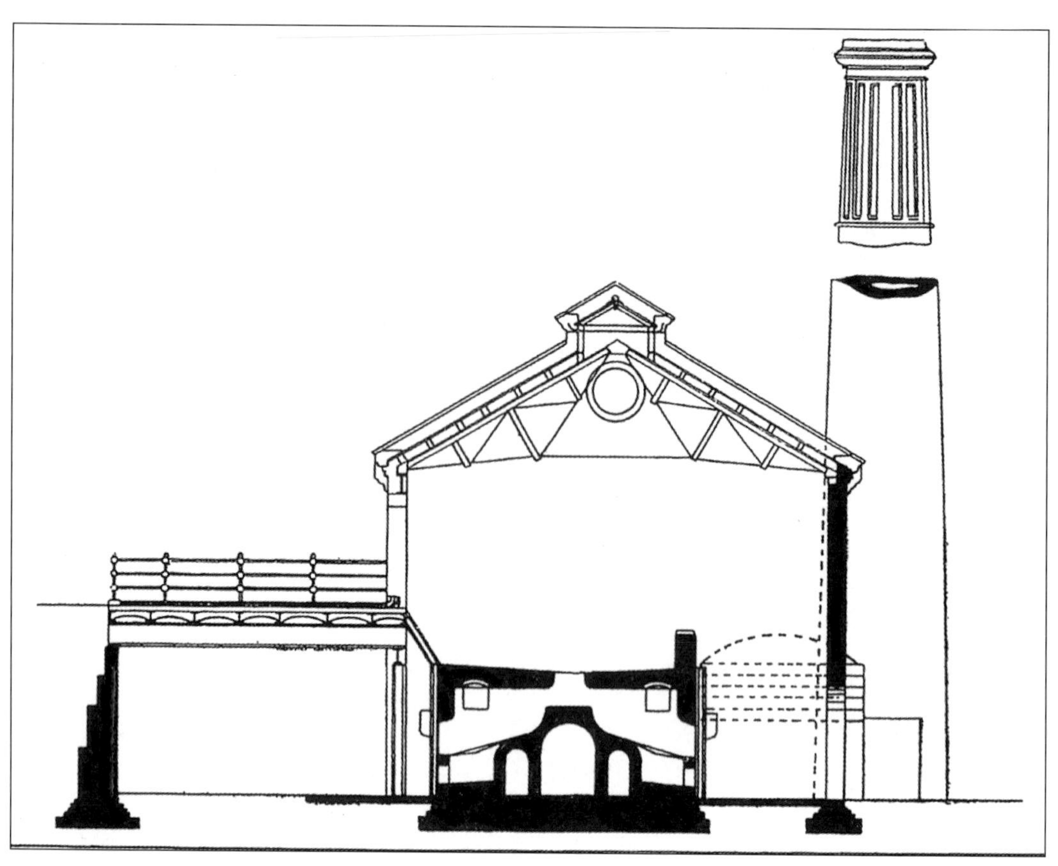

Above:
Cross-section of the Cairo four-cell top incinerator. Built in 1904, it could process 40 tons per day. *Bernard Titcombe collection*

Right:
Section of four-cell 'tub-feed' plant built by the Horsfall Destructor Co Ltd. *Bernard Titcombe collection*

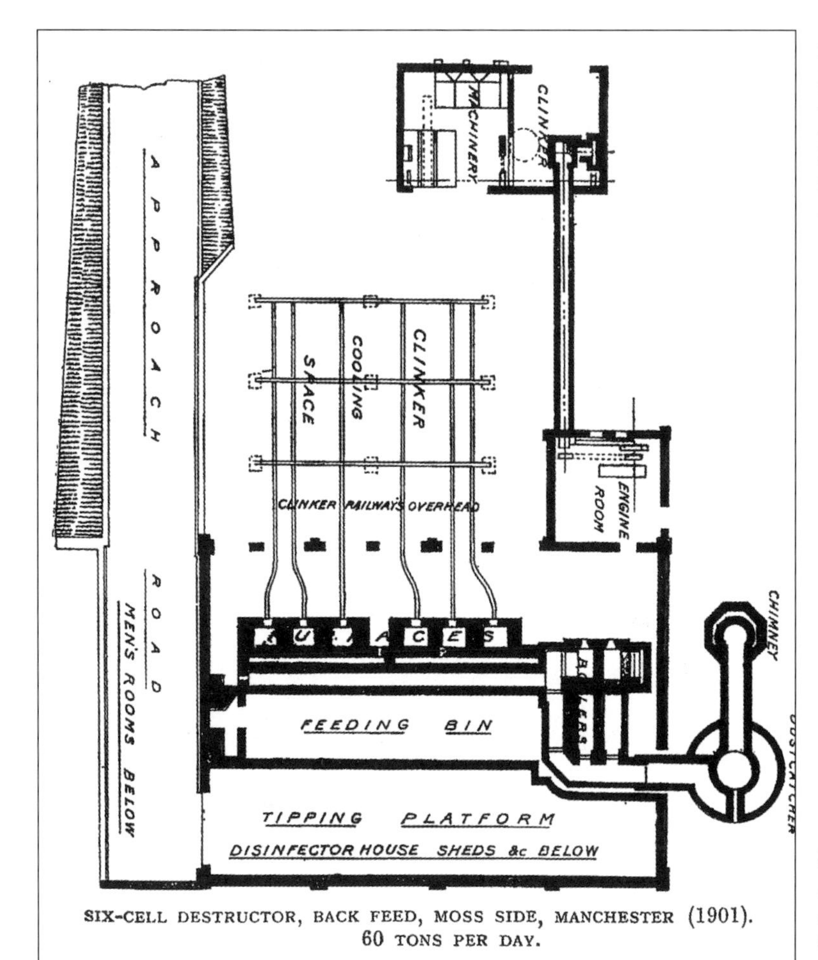

SIX-CELL DESTRUCTOR, BACK FEED, MOSS SIDE, MANCHESTER (1901).
60 TONS PER DAY.

Left:
Diagram showing a back-feed six-cell destructor, which could process 60 tons per day. The plant was built at Moss Side, Manchester in 1901.
Bernard Titcombe collection

Below:
The Horsfall 'tub-feed' destructor charging floor, showing crane, gear and storage tubs.
Bernard Titcombe collection

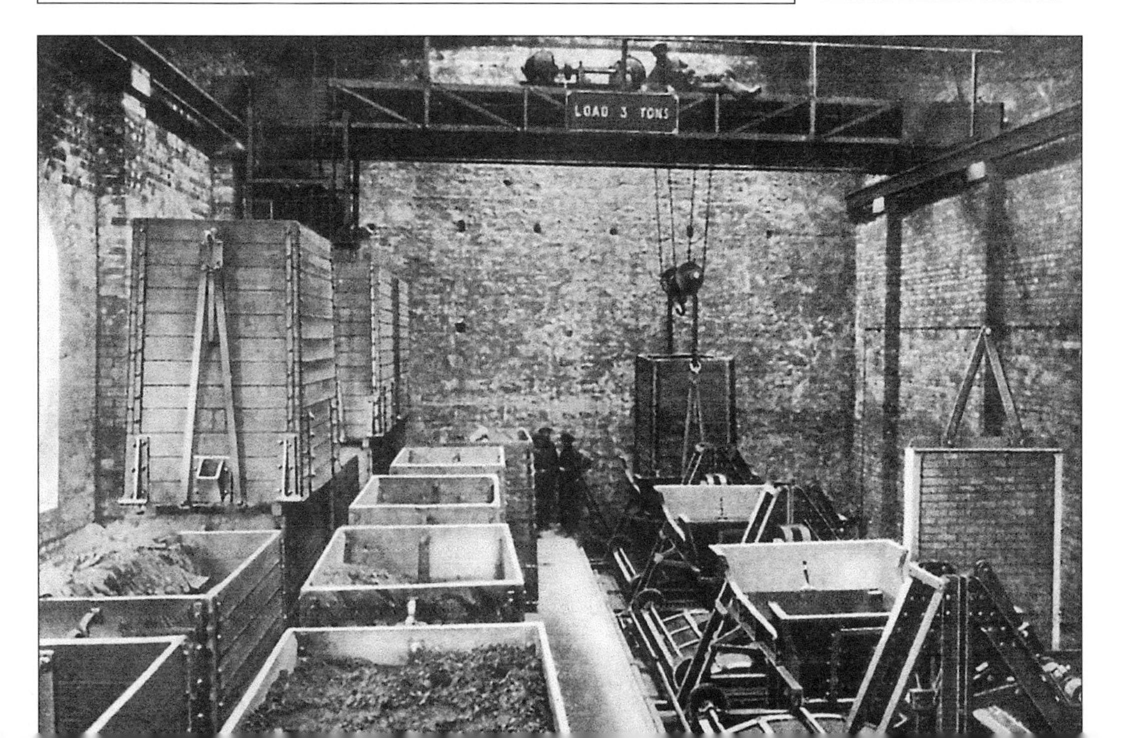

Above: The Heenan refuse destructor (patented 1906/1910) with its 'improved clinkering arrangement by means of a trough gate'. *Bernard Titcombe collection*

Above: The Meldrum destructor was used in many towns and cities. The original caption claims that a special feature was its 'special offal charging shoot in conjunction with a standard front-feed'. *Bernard Titcombe collection*

Above: **The Stirling destructor at Sydney in Australia.** *Bernard Titcombe collection*

Above: **A view of clinkering at the Warner destructor (location unknown).** *Bernard Titcombe collection*

Above: **Baker's refuse destructor at Phoenix Wharf, used by Finsbury Borough Council.** *Bernard Titcombe collection*

Right: **Baker's patent semi-portable hospital destructor No 1, complete with forced draught fan and electric motor.** *Bernard Titcombe collection*

CUSTODIS CHIMNEYS

have been erected in connection with the following
DESTRUCTOR and SEWAGE PLANTS:—

Ayr, Acton, Aberdare, Blackpool, Bedford, Birmingham, Barnes, Bradford, Clydebank, Chiswick, Cov'try (2), Coombe, Dorking, Droylsden, Ealing, East Ham, Greenock, Galashields, Hanley, Holyhead, Kings Norton, Kingstown, Mansfield, Malvern, Poplar, Paignton, Portsm'th (2), Staleybridge, St. Albans, Stoke New'ton, Surbiton, Stone, Taunton, Yroon, Wimb don (2), Windsor, Wrexham, Walthamstow, Wood Green, Wallasey, Watford

Write for Particulars, Illustrated Pamphlets, Advice & Estimates:
ALPHONS CUSTODIS CHIMNEY CONSTRUCTION Co. LTD.,
119, Victoria Street, Westminster, S.W.
Tel: 779 Victoria. Telg: Custodomus.

Above:
Three Babcock & Wilcox boilers, each with 1,827sq ft heating surface and fitted with superheaters. They were fired with waste heat from Meldrum of Woolwich's patent simplex refuse destructors.
Bernard Titcombe collection

Left:
Advertisement for Custodis Chimneys, which constructed the chimneys at many of the early destructor and sewage plants.
Bernard Titcombe collection

MOTORISED WASTE COLLECTION AND CLEANSING TO THE PRESENT DAY

During the 20th century motorised carts and street sweeping machines gradually took over from the horse and cart. Coincidentally, the construction of the smooth new road surfaces made for easier and more efficient cleaning of roads and paved areas. However, in the early years of the 20th century a large quantity of animal dung was still being deposited on the roads and the quicker it was cleaned off the better, as it could soon disintegrate the tarmac surface. Manufacturers vied with each other to draw the attention of early road contractors and councils to the availability of suitable tarmac for the pavements, streets and roads during the era when horse-power was about to give way to motorised transport.

The sheer number of manufacturers of refuse collection vehicles and street cleaning machines is astonishing. Many versions were made by the manufacturers of the vehicle, while others were designed and manufactured to fit on to the chassis of vehicles manufactured by the likes of Leyland, Ford, Seddon-Atkinson and others. Some manufacturers, for example Dennis, not only made the chassis but the refuse container as well, as did Faun, and Pagefield.

After the interruption of World War 2 many of the early manufacturers of waste disposal and street cleansing vehicles continued their production lines. Today some of these names have disappeared, other companies such as Dennis are flourishing, and new names, many from abroad, such as Scania, have appeared on the scene.

Above: **The age of steam is upon us, with this Mann's steam watering wagon.** *Bernard Titcombe collection*

Left:
A petrol-powered watering wagon from Thornycroft.
Bernard Titcombe collection

Above:
The city of Paris' motor street sweeping machines in 1900.
Bernard Titcombe collection

Left:
Thomas Green was one of the early manufacturers of motorised road rollers. Here, however, Green turned his hand to producing motorised street sweeping machines.
Bernard Titcombe collection

Above: One of the early four-cylinder petrol wagons, seen here at Westminster working for the city council. Its engine developed 30hp. *Bernard Titcombe collection*

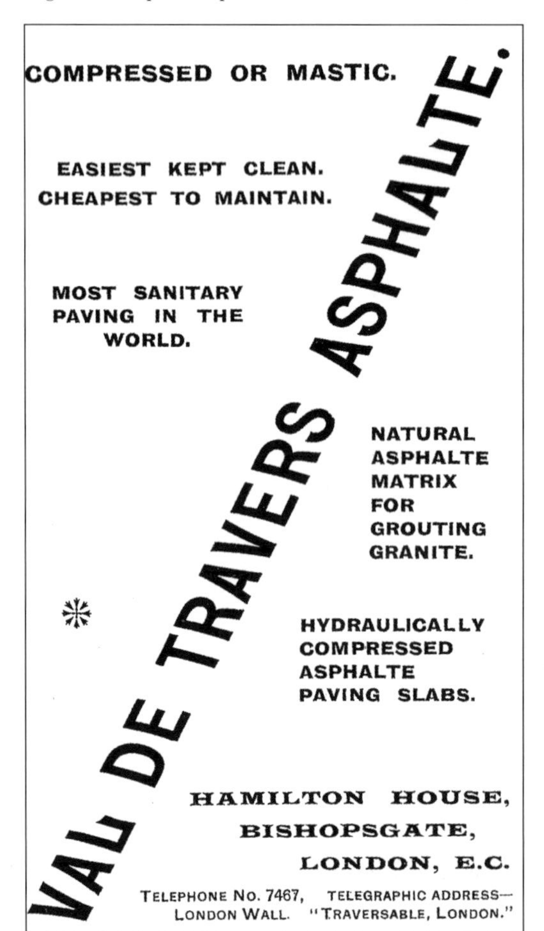

Above: An advertisement from one of the major suppliers of asphalt. *Bernard Titcombe collection*

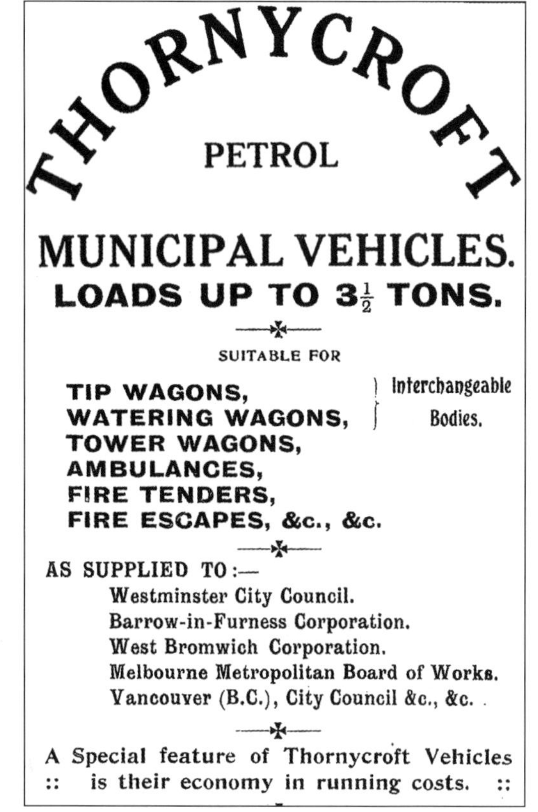

Above: A major manufacturer of lorries, including municipal vehicles, was John I. Thornycroft & Co Ltd of Westminster, London. Its trading success took up almost the whole of the 20th century. This advertisement dates from the early years of the last century. *Bernard Titcombe collection*

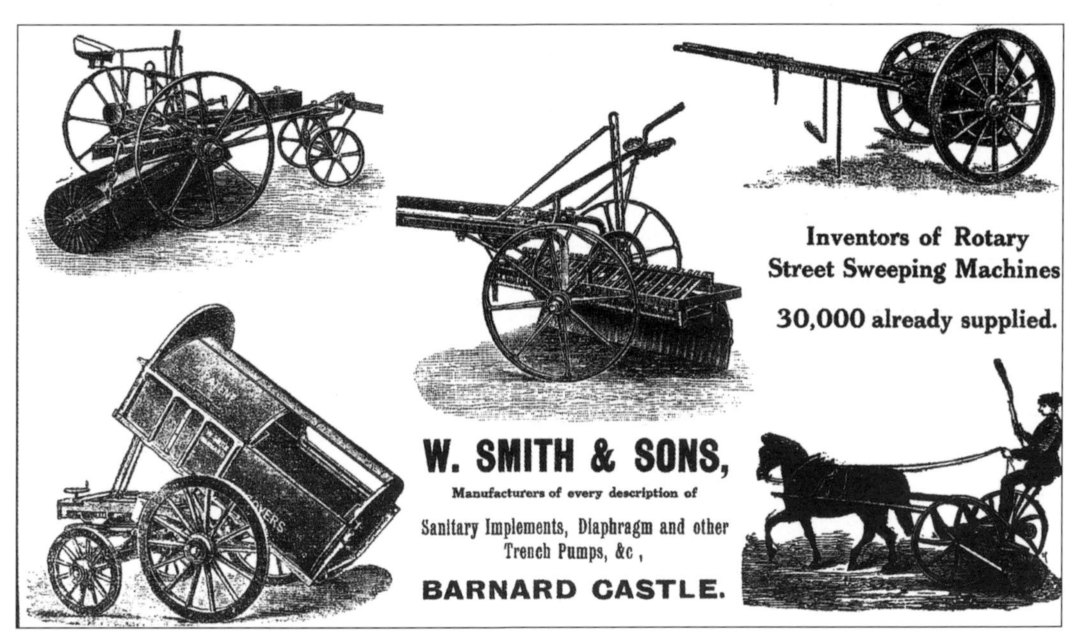

Above: W. Smith & Sons advertisement for the manufacture of horse-drawn implements, carts and wagons. *Bernard Titcombe collection*

Above: Watering and flushing carts and vans, made by Maldon Iron Works Co Ltd. *Bernard Titcombe collection*

The Improved Wood Pavement Company's

CREOSOTED DEAL PAVING

THE MOST DURABLE AND MOST SANITARY of all WOOD PAVEMENTS. Gives a SMOOTH AND EVEN SURFACE for a LONGER PERIOD than other Pavements.

Aldwych
Aldersgate Street
Aldgate
Bayswater Road
Buckingham Palace Road
Brixton and Herne Hill (tram tracks and sides)
Bishopsgate
Bow Street
Bedford Road
Brompton Road
Broad Sanctuary
Brockley Road
Bond Street (New and Old)
Charing Cross
Chapham Road
Charing Cross Rd.
Chiswick High Rd.
Clerkenwell Road
Chalk Farm Road (tram tracks)
Coldharbour Lane (tram tracks and sides)
Chiswell Street
Chancery Lane
Constitution Hill
Castlenau Barnes
Crutched Friars
Earl's Court Road
Edgware Road
Eastcheap.
Fleet Street
Finsbury Circus
Finchley Road
Fulham Road
Golden Lane
Gloucester Road
Grosvenor Place
Hampstead Road (tram tracks)
Hyde Park Corner
Haymarket
Herne Hill Road (tram tracks and sides)
High Street, Notting Hill
Kensington High St.
Kensington Road
Knightsbridge
Kennington Park Rd
King's Cross Station Approach
Kingsway
King William Street, E.C.
Lambeth Palace Rd.
Leicester Square
Ludgate Circus
Lewisham High Rd.
Loughborough to Norwood (tram tracks and sides)
Lower Kennington Lane
Marble Arch Improvement
Milkwood Road (tram tracks and sides)
Mitchan Lane (tram tracks)
Northumberland Ave.
New Bridge Street
Norwood Rd. (tram tracks and sides)
Old Bailey
Piccadilly
Putney to Wandsworth (tram tracks)
Parliament Street
Park Lane
Pimlico Road
Queen Victoria St.
Queen's Road
Regent Street
Richmond main streets
Strand
Shaftesbury Avenue
St. James's Street
St. Martin's Lane
St. Paul's Churchyard
Smithfield
Shepherd's Bush Rd.
Tooley Street and Deptford (tram tracks)
Tower Bridge Approach
Uxbridge Rd., A
Upper Richmond
Vauxhall Bridge
Westbourne Gr
Westminster Br
Whitehall
Wellington Stre
Westminster B Road
Wood Lane, H mersmith

and HUNDREDS of other important thoroughfares in London and the Provinces to the extent of over

350 MILES ALL PAVED BY THE COMPANY.

— THE —

IMPROVED WOOD PAVEMENT COMPAN

— LTD., —

46, Queen Victoria Street, London, E.C.

ESTABLISHED 1872.

Left:
An advertisement for creosoted wooden pavings made from deal. Local authorities in London and the provinces were the customers.
Bernard Titcombe collection

"LEYLAND."

5-Ton Municipal Tip Petrol Motor.

24 H.P. Aubulance Van.

LONDON OFFICES:
47, New Kent Road, S.E.

MAKERS OF :—
Municipal Motors. (Steam or Petrol)
Omnibuses
Tower Waggons
Watering Carts
Police Patrol Vans
Ambulances
Fire Engines
Fire Tenders
Gulley Flushers.
&c., &c.

TIP MOTORS FOR
HighwaysDepartments
Gas Works ,,
Cleansing ,,
Street Watering ,,
&c. &c.

Some Corporations Supplied :—

Chelsea
Westminster
Wandsworth
Twickenham
Liverpool
Blackburn
Pontypridd
Belfast
Barnet
Manchester
Haslingden
Southport
Leicester
Burnley
Atherton
Morecambe
Bury
London
London County Council
Sheffield
Dublin
Birkenhead
Sydney
Green Point
Bombay
Calcutta
Hamburg
Monte Video
Rangoon
Cape Town
Lisbon
Parahyba
&c., &c.

Send for our new Catalogue.

PONTYPRIDD URBAN DISTRICT COUNCIL

6-Ton Municipal Tip Steam Motor.

LEYLAND MOTORS LTD., LEYLAND, LANCS., ENGLAND.

Right:
Leyland Motors, for many years, supplied a wide range of municipal vehicles. Now under the ownership of DAF, little has changed as the current Leyland DAF models still regularly appear in the livery of local authorities and municipal undertakings.
Bernard Titcombe collection

FOR

MUNICIPAL MOTORS

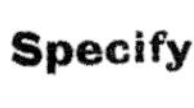

Specify

DUNLOP SOLID TYRES,

As used by

THE LEADING RAILWAY AND MOTOR BUS COMPANIES.

Dunlop Solid Tyres, like the best Industrial Motors, are

BRITISH MADE.

DUNLOP RUBBER COMPANY, LIMITED.

MAN MILL BIRMING

Left:
Dunlop solid tyres as advertised for municipal motors.
Bernard Titcombe collection

Green's MOTOR

STREET SWEEPING MACHINES.

Will do the work of at least Three Horse-Sweeping Machines and much more efficiently.

Simple, Durable and Reliable.

Are also makers of Steam Road Rollers, Tractors Vertical and Cornish Boilers.

GREEN'S MOTOR ROLLER for . . Tar Macadam.

Water Ballast or Ordinary Pattern.

FITTED WITH TWO SPEEDS AND REVERSE.

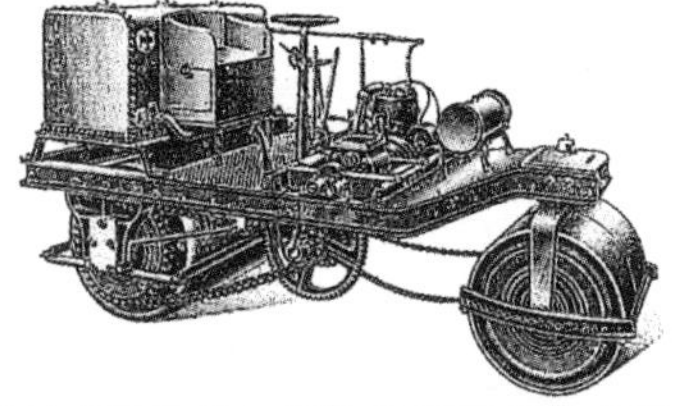

These Motor Rollers are made in sizes from 3 to 10 tons.

THOS. GREEN & SON, LTD.,

Smithfield Iron Works, Leeds and New Surrey Works, Southwark Street, London, S.E.

FULL PARTICULARS ON APPLICATION.

Right:
An advertisement for Green's motor street sweeping machines and motor roller.
Bernard Titcombe collection

Pagefield — A Company History

For this company history I am indebted to Mr Tom Meadows, a life-long employee with the company. He has allowed me to use extracts from his own account of life with Walker Bros of Wigan, the manufacturer of Pagefield vehicles.

Pagefield was originally formed by John Scarisbrick Walker. Born in 1842, he was one of five sons of Mr James Walker, a cordwainer (or shoemaker) who lived near the marketplace in Wigan, and who died in 1856. John attended Wigan Grammar School, as did his four brothers. On leaving school he began an apprenticeship at Haigh Foundry. This was the first foundry in Wigan, built in 1789 by Alexander, Sixth Earl of Balcarres, his brother Robert Lindsey and James Corbett. It was the ideal place to gain a thorough knowledge of engineering, as during the years that John Walker spent there 114 locomotives were produced for the fast-growing railway market at home and abroad. Other engineering products, such as swing bridges for the Albert Dock in Liverpool, cast iron tubing and arches for the Mersey rail tunnel, and sugar machinery for the West Indies, were also produced.

John Walker's main interest lay in colliery requirements, which for the area around Wigan, Coppull, Chorley and Bolton were substantial. He left Haigh Foundry in 1866 at the age of 24 to set up in business on his own manufacturing nuts, bolts and pit tubs in small premises in Queen Street, Wigan. A prize of £500 was being offered by colliery proprietors in South Lancashire for the most suitable design of coal cutter and it is presumed John Walker claimed the prize when his company, and he in particular, designed the world's first mechanical coal-cutting machine. His brother Thomas, who had already had a brief experience as a mining engineer, joined John Walker in the business in 1869.

By the late 1870s, all manner of mining machinery was being built by the Walker brothers, not only for the domestic market but for export, particularly to Belgium. This included air compressors, fans, coal cutters, etc. By 1873 it was realised that the small premises at Queen Street were totally inadequate to house the requirements of the manufacturing facilities and finished products, so five acres, known as Page's Fields, were acquired just out of Wigan. Hence the Pagefield Ironworks was formed and became a Limited Company in 1904.

In the same year (1904) Pagefield Motor Vehicles were introduced, being manufactured alongside the railway locomotives which the company had been building since 1890. (Many of the shunting locomotives were used in the construction of the Manchester Ship Canal.) Although some thought had been given to the production of motor cars, it was felt that the market for this would be too competitive, so it was decided to concentrate on heavier vehicles. One of the first orders was in 1908 for a 'tower' wagon for Rushton of Wigan and Wigan Corporation. Other vehicles sold prior to World War 1 included wagons for Burtonwood Breweries, Wigan Co-op, Small Brook Potteries, W. R. Deakin and petroleum companies such as BP. Two omnibuses were also delivered in 1914. Other customers between 1908 and 1914 included W. & R. Jacob (biscuit manufacturers) and Wigan Coal & Iron Company.

In 1911 2-ton-capacity wagons were fitted with Dormer or Tyler engines, which developed 28bhp at 1,200rpm and consumed a gallon of fuel for every 10 miles. The same engine was used in 3-ton-capacity vehicles. In 1912 Pagefield 3-tonners entered for the War Office subsidy trials and were subsequently built to War Office specifications. The N-type subsidy Pagefield was approved by the War Office and was shown at Olympia in 1913. It used a Dorman subsidy engine, which developed 42bhp at 1,000rpm. By the end of World War 1 a total of 519 Pagefields had been sent into service. Soon after the war, in 1919, a substantial order was received for vehicles for the Schwager Mines in Chile, for nitrate production.

In 1922 the Borough Engineer of Southport asked Walker Bros if they could supply refuse collection vehicles to replace the horse-drawn carts which found it difficult to make the round trip to the tips, which were far out of town, and back again. For its initial venture into this new growth industry, the company designed the 5-ton W model as the basis for its subsequent Pagefield system vehicles. The Pagefield system had, in fact been developed in the days when a horse and cart was used to remove refuse from towns and villages. In Southport, where the Walker family lived, carts were first used which were able to winch a wheel-mounted refuse container onto the back for transportation to a refuse tip. Once there, the wheel-mounted container was unloaded from the cart and a second empty container would be winched on board so as to speed up the turn-around at the tip.

This system was then adapted for use on Pagefield trucks. The tipping body of the lorry incorporated a winch system run from a power take-off connected to the top of the gearbox. This enabled wheel-mounted containers to be pulled on and off the lorry bed. The tipping device was operated by the same method. The 12cu yd-capacity collection containers were mounted on 20in-diameter wheels. These containers were horse-drawn when collecting refuse, but when full were hoisted on to the lorry which was fitted with a lifting frame instead of a body. The lorry travelled at speed to the disposal point and tipped the load, returning the empty container to a different area, in which another horse-drawn container and a gang of men were operating.

In principle and practice, the Pagefield system utilised two types of vehicles: one designed exclusively for the convenience of loading, whilst the other performed its specialised function of high-speed transport and the negotiating of rough ground on refuse tips.

One London contractor (Surridges) took over the refuse collection and disposal in the Boroughs with a population exceeding one million. In later years, after World War 2, another London contractor (Drinkwater's) took over some of the old Pagefields. From being run by private contractors in those early years, the wheel seems now to have turned full circle — from private to public and now back to private again.

The 1920s and 1930s saw Walker Bros introduce a number of significant Pagefield vehicle types into the refuse collection vehicle market, including the Pagefield Prodigy, the mid-1930s Pagefield Paragon which featured a four-cylinder 25/50hp Meadows petrol engine, or as an option, a Perkins oil engine, and the Pagefield Paladin which was produced from 1937-8 (although the name 'Paladin' had previously been used on another vehicle). The design of the Paladin, featuring a hydraulic ram to lift and empty heavy bins before returning them to a wheeled trolley on the ground, became an industry standard and the design is still in use today.

Despite building a vast array of waste disposal vehicles, as well as charabancs, a wide variety of vans and lorries, railway and mining equipment, and even a range of mobile cranes — including many which found favour with British Rail, Board and Paper Mills — the company sadly ceased producing vehicles in the late 1950s.

Left:
One of the very early Pagefield refuse vehicles built by Walker Bros of Wigan was especially popular with municipal corporations such as, in this case, Liverpool, because being so narrow it could negotiate the very narrow streets and back alleys, without causing damage to property. *Tom Meadows*

Right:
The unique design of this Pagefield refuse vehicle from 1922 has certainly provoked a lot of interest in more recent years, with many current manufacturers applying the roll-on, roll-off technology to today's removal of waste.
Tom Meadows

Above: **These pictures demonstrate the Pagefield system.** *Tom Meadows*

Above: The first Pagefield system vehicle fitted with the Gardner 4L2 diesel engine. Almost everything, including the ash cab, was designed and produced at Pagefield. *Tom Meadows*

Above: Another Pagefield system vehicle, being used as a spare vehicle for road spraying by St Pancras Borough Council. *Tom Meadows*

Right:
Large numbers of these Pagefield Prodigy refuse collectors were built and most councils used them as a very adaptable back-up vehicle.
Tom Meadows

Left:
The London Borough of Holborn specified a complete metal body for all of its Pagefield Prodigy refuse collection vehicles.
Tom Meadows

Right:
The County Borough of Wigan (Streets Department) was among the many authorities to use the general-purpose Prodigy from Pagefield.
Tom Meadows

Above: A Pagefield Prodigy, equipped with a 750-gallon tank for road maintenance work was being used by St Marylebone Borough Council. Note its sprinkler system with a nearside flushing pipe and a rear-mounted sprinkler, which had a spread width of around 20ft. *Tom Meadows*

Above: Lancaster Corporation used a 10cu yd body on its canopy-type Pagefield refuse vehicle. *Tom Meadows*

Right:
The Pagefield Paragon was introduced in the mid-1930s. It featured a Meadows petrol engine of 25/50hp via its four cylinders, or as an option, a Perkins oil engine. *Tom Meadows*

Left and below:
The rear or work end of the Pagefield Paragon. *Tom Meadows*

Above:
The Pagefield Paladin became an industry standard, and the design is still much in use today.
Tom Meadows

Left:
The Pagefield Paladin was produced from 1937-8. It was designed primarily to ease the unloading of heavy bins, often used for large blocks of flats, and made use of a simple hydraulic ram to perform the lifting and emptying process before returning the empty bin to the ground, where the small-wheeled trolley was placed to roll the bin back to its original position. *Tom Meadows*

Left:
This shows the hoisting and discharge method in use on the Paladin. *Tom Meadows*

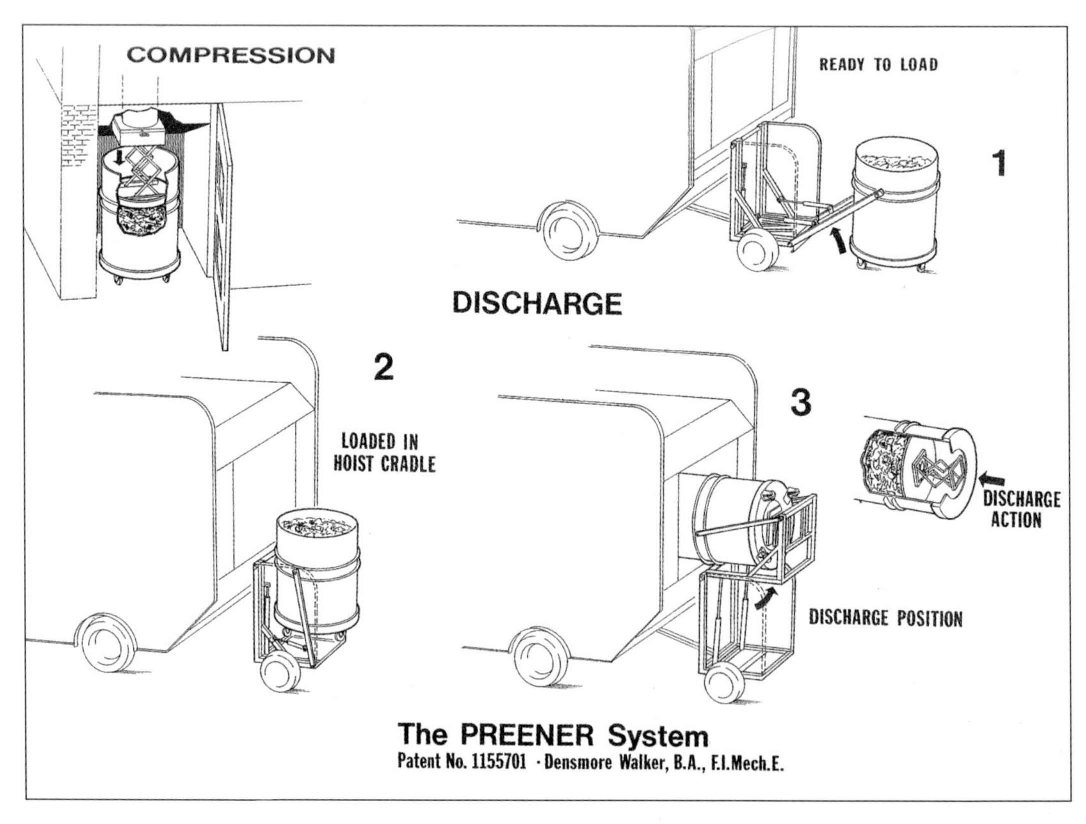

Below:
The Preener System, patented by Walker Bros, is demonstrated in this diagram. *Tom Meadows*

Left:
One of the very earliest refuse vehicles from 1929, from General Vehicles of London. *Ian Allan Library*

Right:
An Electricar refuse vehicle at work in the autumn of 1931. *Ian Allan Library*

Left:
Another version of the General Vehicle Co's refuse vehicles in June 1933. *Ian Allan Library*

Left:
A General Vehicle electrically-powered refuse wagon in Barrow-on-Furness Corporation's employ in November 1934.
Ian Allan Library

Right:
A Morris-Commercial 7cu yd capacity refuse collector in November 1934.
Ian Allan Library

Left:
A Morris Commercial refuse collector at work in the parish of St Helier in December 1938.
Ian Allan Library

Right:
A Fordson 7cu yd capacity refuse vehicle in 1936.
Ian Allan Library

Left:
Wortley Rural District Council was using this 2-ton forward-control-chassis Fordson in March 1938.
Ian Allan Library

Right:
A Vulcan refuse collector seen on 12 June 1937.
Ian Allan Library

Above:
A Dennis 10cu yd side-loading refuse collector with an ARP trailer in November 1938.
Ian Allan Library

Right:
A Dennis Compactum machine at work for the City of Portsmouth in March 1934.
Ian Allan Library

Left:
This 7cu yd Dennis refuse vehicle is from June 1937.
Ian Allan Library

Above: This little Dennis refuse vehicle seen in 1932 had the use of full hydraulic tipping, even then.
Ian Allan Library

Above: This 7cu yd Dennis was chosen by a number of councils, seen here in December 1937 demonstrating its high angled tip. *Ian Allan Library*

Above: The Dennis Handyman, shown here in 1931, equipped as a gully sucker/sewerage disposal vehicle for the City of Coventry. *Ian Allan Library*

Above: This refuse vehicle from June 1939 was used to carry a water tank and hose system. *Ian Allan Library*

Left: These Thornycroft 'Handy' 2-ton-capacity vehicles were at work in September 1935. The refuse collector had a capacity of 7cu yd. *Ian Allan Library*

Right: An 8cu yd-capacity refuse collection vehicle incorporating the Thornycroft Handy class 2-ton chassis and an all-steel body with three sliding aluminium covers on each side. They were supplied as a repeat order for Durham Rural District Council in June 1937. *Ian Allan Library*

Left: This large refuse van was produced by John L. Thornycroft in June 1934. *Ian Allan Library*

Above: John L. Thornycroft supplied this side-loading refuse collector in June 1938 to the County Council of Lanark, Bellshill & Mossend Special District. *Ian Allan Library*

Above: A Bedford refuse collection vehicle, with its side-loading compartments. The date is unclear, but probably the late 1920s/early 1930s. *Ian Allan Library*

Right:
A 7cu yd-capacity Albion in November 1932 at work receiving refuse into its side-loading hoppers.
Ian Allan Library

Left:
An Associated Equipment Co (AEC) petrol-engined Monarch with an Ideal refuse collector and its container in the travelling position, in June 1938.
Ian Allan Library

Right:
A Vulcan refuse collector from June 1937. It features a 2-ton low-loading chassis fitted with rear-loading barrier-type refuse body, with a moveable floor. Vulcan was based in Southport, Lancashire.
Ian Allan Library

Above: An 18/30cu yd Compressmore refuse collection vehicle incorporating the Thornycroft Trusty chassis and Eagle body with patented hydraulically-operated compressing device in June 1937. *Ian Allan Library*

Above: A Dennis Rafidain cesspool emptier, as used by local authorities, in July 1931. *Ian Allan Library*

Above: A high pressure water jet is used for street washing. This one, in June 1937, was mounted for use on a Scammell. *Ian Allan Library*

Left:
This Dennis refuse collector was in use for the City of Oxford in August 1937.
Ian Allan Library

Right:
A mechanical, motorised roadsweeper at the Water Street Depot of Manchester Corporation Cleansing Department seen on 26 September 1928.
Mike Pilkington, Manchester City Council

Left:
A Karrier refuse carrier from the Water Street Depot of Manchester Corporation Cleansing Department at work on 16 September 1928.
Mike Pilkington, Manchester City Council

Above: A Shelvoke & Drewry gully cleaner at work on the streets of Manchester on 1 April 1938.
Mike Pilkington, Manchester City Council

Above: A large mechanical brush at work in Manchester on 1 April 1938. *Mike Pilkington, Manchester City Council*

Above: The County Borough of East Ham Public Cleansing Department ordered 12 of these refuse collectors mounted on Pony trucks, built by Brush Coachworks Ltd of Loughborough in August 1947. *Ian Allan Library*

Above: Battery Traction Ltd was the concessionaire for this refuse vehicle made by Bleichert of Leipzig, Germany. It featured a unique two-way tipping system mounted on a special low-loading chassis (date unclear, probably mid-1930s). *Ian Allan Library*

Above: The Hygienurba dust cart, being tested for the City of Paris in 1934. *Ian Allan Library*

Above: Another look at the Hygienurba, at work in the streets of Paris in 1934. *Ian Allan Library*

Above:
A Scammell mechanical horse at work in January 1938 utilising a complete tender to deal with poison gas contamination. A squad of up to 12 men were trained to work with this vehicle.
Ian Allan Library

Left:
This picture shows the work-end of a Dennis refuse collection vehicle on 21 June 1947.
Ian Allan Library

Left:
A Scammell gully emptier in June 1939, helping Leeds City Council keep its streets clean and water flowing. *Ian Allan Library*

Left:
A Scammell vertical-type bin carrier with motive unit, as supplied to Kensington Borough Council, Battersea Borough Council and others, seen in June 1937. *Ian Allan Library*

Left:
A Scammell 3-ton mechanical horse motive unit with a 1,250-gallon street washing carrier, in October 1937. *Ian Allan Library*

Right:
Another Scammell mechanical horse, hauling a street washing water tank, hoses etc, in June 1948. *Ian Allan Library*

Left:
A Scammell mechanical horse from June 1934 hauling a tar boiler tank made by John Fowler of Leeds. *Ian Allan Library*

Right:
The City of Westminster proudly shows off the portcullis on the coat of arms on this Scammell water tank in February 1941. *Ian Allan Library*

Above: A Scammell road sweeper/collector coupled to the Scarab mechanical horse in the County Borough of Barnsley in the late 1940s. *Ian Allan Library*

Above: A Seddon refuse vehicle ready to go on duty, possibly in the Netherlands, in May 1960. Note the wheel on the outside of the cab. *Ian Allan Library*

Right:
A rear view of the Scammell road sweeper and driver at work in May 1939.
Ian Allan Library

Below:
A Scammell Scarab coupled to an 18cu yd-capacity moving-floor refuse collector, in June 1949.
Ian Allan Library

Right:
A tipper body mounted on a 25/30cwt-capacity Brush battery electric chassis provides the Wexford Corporation Cleansing Department with an efficient and hygienic vehicle for the collection of kitchen waste, in April 1948.
Ian Allan Library

Left:
This Ford Motor Co lorry is fitted with a 12cu yd Derby-type refuse loader made by Eagle Engineering Co Ltd of Warwick. The chassis is a Ford Thames, from July 1947.
Ian Allan Library

Right:
A Ford Thames, fitted with a 500gal-capacity gully and cesspit emptier tank in July 1947. *Ian Allan Library*

Above: This Thornycroft Nippy Star oil-engined house refuse collection vehicle (940 AA), with a Chelsea-type body having four sliding shutters each side and Eagle end tipping gear, was delivered to St Austell Rural District Council in Cornwall in October 1956. Standing alongside is an earlier Thornycroft Nippy 3-ton refuse collector, which the new machine replaced after 17 years of service in the china clay-producing town of St Austell. *Ian Allan Library*

Right: A Commer Karrier CK3, with a 12/15cu yd-capacity moving-floor refuse body and a six-seater cab, is seen in June 1946. *Ian Allan Library*

Right: A Commer Karrier Gamecock 10cu yd-capacity refuse collector with the new TS3 diesel engine, synchromesh gearbox, strengthened rear axle and an improved cab, in September 1954.

Above: A Thornycroft Nippy refuse collection body on the Nippy 10ft 1½in-wheelbase chassis. The Edwards Bros body is of wooden construction, having three sliding covers each side. The vehicle is seen working for the Borough of Buxton, Public Health Department, in Derbyshire; the date is unknown. *Ian Allan Library*

Above: A Commer street wagon working for Gateshead Corporation in February 1945. *Ian Allan Library*

Above: A Commer Karrier Yorkshire, fitted with a 750-gallon cesspool emptier and a night soil tank on the CK3 chassis. It is seen in March 1947 working for Sutton-in-Ashfield Urban District Council. *Ian Allan Library*

Left: A Commer refuse vehicle with enlarged cab for its personnel, photographed circa 1952. *Ian Allan Library*

Left: A Karrier Bantam from Commer with a 7cu yd-capacity refuse collector, powered by a 48bhp engine, seen in July 1955. *Ian Allan Library*

Above: A Commer diesel lorry fitted with a water tank, photographed in March 1941. *Ian Allan Library*

Right: A Commer Karrier Yorkshire 800-gallon gully emptier, with a 91bhp six-cylinder petrol engine. The gully-emptying equipment is of 'in-frame' construction with a vacuum pump driven from the main engine and gearbox. It is mounted here in this 1954 photograph on the ¾-ton chassis, though a 5-ton chassis was also available. *Ian Allan Library*

Right: Rootes Motors, which produced the Commer range of vehicles, built this Commer Karrier refuse collector, complete with refuse sack trailer, for the London Borough of Barking. *Ian Allan Library*

Right: The Karrier Transport Loadmaster compressing refuse collector, fitted with 105bhp TS3 diesel engine, had a capacity of 20-25cu yd and is seen working here for West Ham Corporation's Engineers Department. *Ian Allan Library*

Left: Another Karrier vacuum road sweeper with the body fabricated by the Yorkshire Patent Steam Wagon Co, working here for the Works Department of Hemel Hempstead Borough Council in Hertfordshire. *Ian Allan Library*

Right: This Karrier chassis, seen in November 1962, is employing a 50cu yd continuous loading compression refuse collector on a 13ft 6in chassis, with bodywork by Glover, Webb & Liversidge Ltd. *Ian Allan Library*

Right:
The Royal Borough of Edinburgh was the owner of this street washer built on a 105bhp Rootes diesel-engined 7-ton Commer chassis by the Yorkshire Patent Steam Wagon Co. The tank had a capacity of 1,400 gallons and the centrifugal pump, driven by the power take-off on the gearbox, could maintain a flow of 100 gallons per minute through the delivery valve which was controlled from the driver's cab. The tank had a fully-opening rear door to facilitate easy cleaning and maintenance. Built in February 1958, it was state of the art. *Ian Allan Library*

Left:
The City of Westminster Council's Public Cleansing & Transport Department converted many of its utility and refuse collection vehicles to battery-electric power, as demonstrated by this Dennis Pax Model IV from August 1962. *Ian Allan Library*

Right:
The Karrier Bantam, a 7cu yd refuse collector from May 1952. *Ian Allan Library*

Right:
A Karrier Motors mechanical road sweeper seen in June 1950. *Ian Allan Library*

Above:
An Electricar gully emptier, seen working for the City of Westminster Council in May 1936. *Ian Allan Library*

Right:
This is a Karrier Yorkshire 750-gallon CK3 gully emptier, which was commissioned by Sunderland Corporation (1949). *Ian Allan Library*

Right:
A Karrier CK3 operating as one of Biggleswade Urban District Council's fleet in 1946.
Ian Allan Library

Left:
A Karrier Bantam 7cu yd refuse collector with a side-loading, steel-lined body, and hydraulic tip, semicircular sliding covers and double rear doors, in service for Warmley RDC, photographed in July 1947.
Ian Allan Library

Right:
A 1948 Bantam 7cu yd steel-lined refuse collector from Karrier, seen working in Stoke Newington.
Ian Allan Library

Above: This Karrier was one of many which were exported, this time to Brussels in 1950. *Ian Allan Library*

Above: Another Karrier export is seen equipped for refuse collection in the Municipality of Gatooma (Zimbabwe), in October 1950. *Ian Allan Library*

Right:
This odd-looking vehicle was a Dennis Vulture forward-loading refuse collection lorry built in July 1953.
Ian Allan Library

Right:
A Dennis refuse collection vehicle was an important exhibit in the Lord Mayor's Show in 1956.
Ian Allan Library

Left:
A 2-ton Karrier Bantam refuse collector at Hetton Urban District Council in July 1944.
Ian Allan Library

Above: Bath City Council used this Albion refuse wagon with hydraulic end-tipping gear in October 1941. *Ian Allan Library*

Above: A 17cu yd-capacity Ford refuse collector in June 1949. *Ian Allan Library*

Left:
A Thornycroft van from June 1948 at work in the Borough of Willesden for its Cleansing Department.
Ian Allan Library

Right:
An 800-gallon gully emptier mounted on a Thornycroft Sturdy, working for the Sutton-in-Ashfield Urban District Council Surveyor's Department.
Ian Allan Library

Left:
A Dennis refuse collector, seen in July 1947.
Ian Allan Library

Above:
On demonstration in September 1945 is this Dennis gully emptier with its 800-gallon tank, at work for the Cleansing Department of Blackpool Corporation.
Ian Allan Library

Right:
This Dennis van, seen in June 1949, features open-out windows and a handle start.
Ian Allan Library

Right:
A Dennis 750-gallon gully emptier at work for the Borough of Hammersmith in 1945.
Ian Allan Library

Below:
Aylesbury Rural District Council used this Dennis gully emptier during a demonstration in September 1945.
Ian Allan Library

Left:
This 12cu yd-capacity refuse collection vehicle from Dennis was on its way in September 1945 to the Borough of East Ham. *Ian Allan Library*

Above: This Dennis Paxit II refuse collection vehicle, was powered by a Perkins P6 diesel engine rated at 83bhp at 2,400rpm, and was one of seven shipped to the Colombo Municipality of Ceylon (Sri Lanka) in December 1961. *Ian Allan Library*

Above: In the West End of London a Dennis Pax IV articulated refuse collector negotiates the turning off Curzon Street into Clarges Street in February 1963. *Ian Allan Library*

Left:
The most ubiquitous bodybuilder at the 1966 Commercial Motor Show at Earls Court, London was Edbro Ltd of Bolton which built the tipping gear and body for this refuse collector. The vehicle chassis, built by Austin Crompton Parkinson Electric Vehicles Ltd, was one of 28 bought by the Newham Borough Council. It had a capacity of 3cu yd. Tipping was undertaken by a single hydraulic ram with tipping gear power-driven from an electrically-operated pump. *Ian Allan Library*

Right:
This Karrier Yorkshire 750-gallon gully emptier was powered by a Perkins P6 diesel engine when produced in 1953. *Ian Allan Library*

Below:
This refuse collection vehicle was in service with Beesel Municipality in the Netherlands. The bodywork was by N. V. Buca of Buchten, on the 179in-wheelbase 7-ton chassis powered by the Bedford 330cu in diesel engine. *Ian Allan Library*

Above:
One of the earliest of all mechanical road sweepers was this Elgin from 1910.
Johnston Engineering Ltd

Right:
An Elgin street washer in 1920.
Johnston Engineering Ltd

Above: A three-wheeled mechanical sweeper from Elgin at work in 1914. *Johnston Engineering Ltd*

Above: Spot the difference? An Elgin Pelican three-wheel mechanical sweeper from the present day. *Johnston Engineering Ltd*

Above:
An early JEL mechanical sweeper from 1937. *Johnston Engineering Ltd*

Right:
An advertisement for a Johnston suction cleaner from the 1950s.
Johnston Engineering Ltd

Above: FMC presents here its three-wheeled mechanical sweeper. The date is probably circa 1984.
Johnston Engineering Ltd

Above: The Johnston Sweeper Co V4000 four-wheel mechanical sweeper, built for California Transport (Caltrans).
Johnston Engineering Ltd

JOHNSTON

Left:
A Johnston V4000 in New York City.
Johnston Engineering Ltd

- The new, very compact sweeper includes a built-in wanderhose for quick gully emptying.
- Die neue, sehr kompakt gebaute Kehrmaschine ist mit einem schwenkbaren Schlauch zum raschen Entleeren von Straßen-Gullys ausgerüstet.
- La nouvelle et très compacte balayeuse mécanique comprend un tuyau orientable d'aspiration incorporé pour vidange rapide du caniveau.
- La nuova compatta spazzatrice con incorporato il tubo movibile per svuotare rapidamente le fognature.
- Den nya, mycket kompakta sopmaskinen har också en inbyggd slang för snabb slamtömning.

Melford Sweeper

The new British Melford Precinct is a small, compact and very manoeuvrable ride-on road sweeper offering a hopper capacity of 1.3 cu. yd. (1 cu. m.) and a sweeping system said to be capable of dealing with 90° internal or external corners in one pass.

It is powered by a rear-mounted Ford 2264 E 1600 cc 4-cylinder petrol engine with hydrostatic drive to the single-steered front wheel. Travel speeds of up to 30 mph (50 km/h) are designed to enable the unit to keep up with the traffic stream while the automatic transmission ensures a smooth sweeping action up to a maximum of 7.5 mph (12 km/h). The sweeper width varies between 4 ft. 11 in. (1500 mm) and 5 ft. 11 in. (1800 mm). Fully laden, according to Melford, the Precinct will sweep up a 25% gradient and, with a tipping height of 4ft. 3 in. (1300 mm), can empty direct into standard skips when required. It also has a built-in wanderhose for quick gully emptying. Other features include rubber front and rear suspension and a comfortable enclosed cab for good visibility and control over the forward mounted brushes. No damping is required; the machine works on the same principle as a domestic vacuum cleaner. An advanced dust control system is designed to keep the atmoshere clean. Mr. Ian Duncan, Melfords Managing Director, tells us that he is now developing a brush with variable sweep for negotiating lamp posts and other obstructions without leaving large unswept areas.

20

Right:
A small concept sweeper advertisement in the 1980s.
Johnston Engineering Ltd

Below:
Mrs Margaret Thatcher, the then Prime Minister, sits at the controls of a Schmidt streetwasher in the 1980s. *Johnston Engineering Ltd*

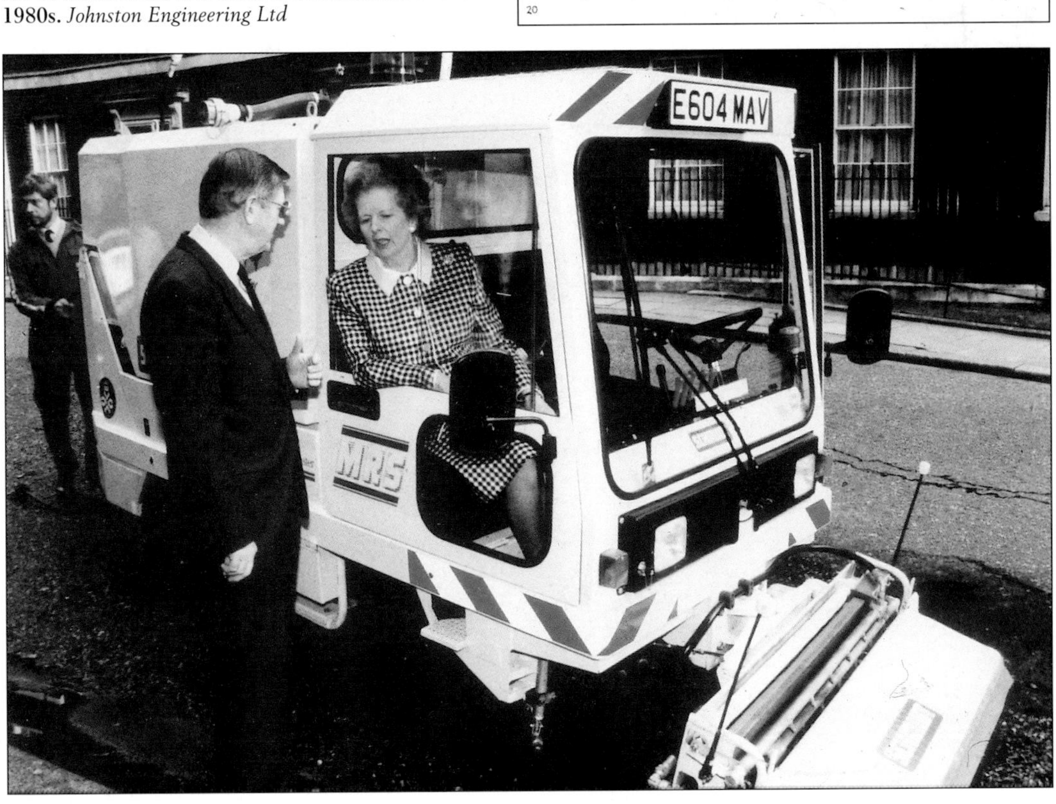

Left:
A new Schmidt being launched in 1998.
Johnston Engineering Ltd

Right:
An early Schörling from the early 1950s.
Johnston Engineering Ltd

Left:
Johnston Engineering units in the 1950s.
Johnston Engineering Ltd

Above: A present-day sweeper from Bucher-Schörling. *Johnston Engineering Ltd*

Above: An air regenerative unit from the American manufacturer Tymco. *Johnston Engineering Ltd*

Right:
An early walk-behind sweeper by Applied Sweepers Limited.
Johnston Engineering Ltd

Below:
Up-to-date Applied Sweepers units.
Johnston Engineering Ltd

Above: An ultra-modern sweeper machine, with bodywork by Renault, at work in the City of London.
Johnston Engineering Ltd

Above: A contractor in the UK using a mechanical sweeper with full-width rear suction nozzles.
Johnston Engineering Ltd

Above: The new 5000 series of sweepers from Johnston Engineering Ltd. *Johnston Engineering Ltd*

Above: Early Elgin Pelican mechanical street sweepers. *Johnston Engineering Ltd*

Above: **A modern-day Applied Sweepers street sweeper.** *Johnston Engineering Ltd*

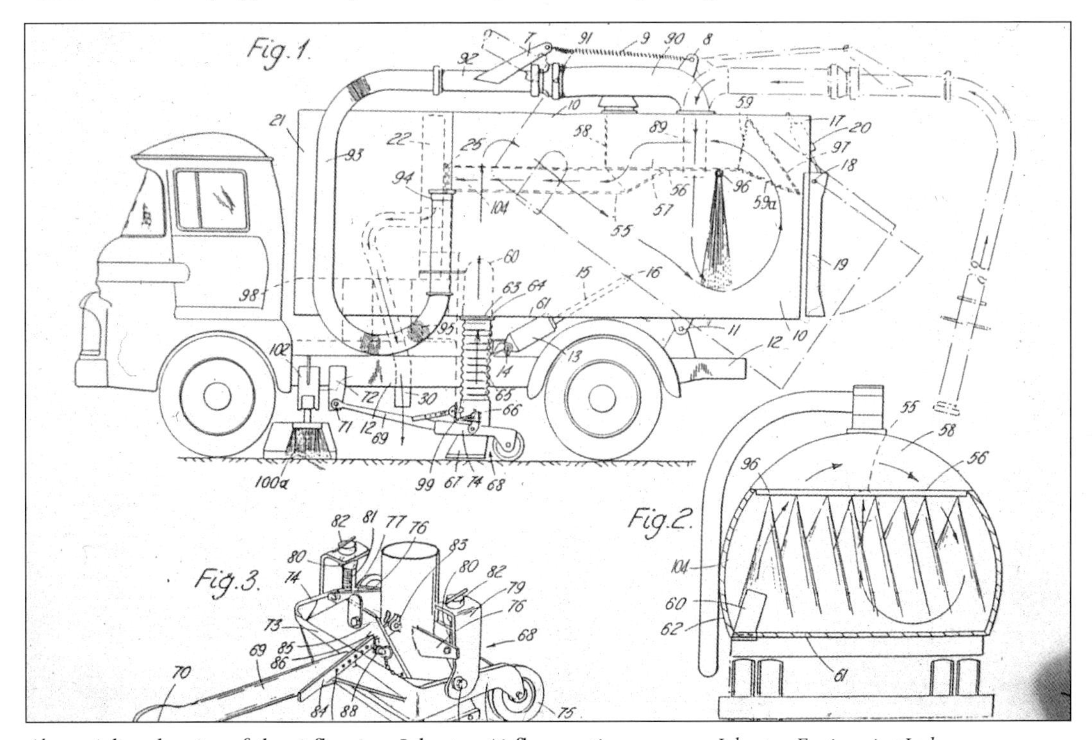

Above: **A line drawing of the airflow in a Johnston Airflow suction sweeper.** *Johnston Engineering Ltd*

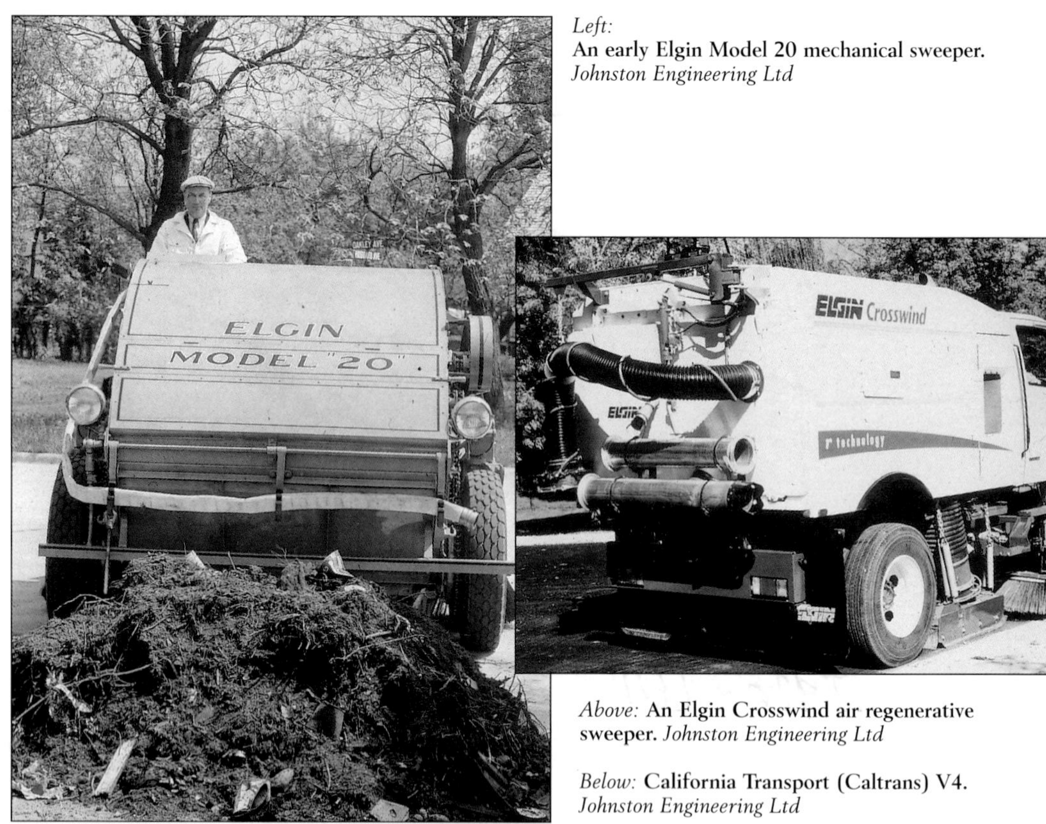

Left:
An early Elgin Model 20 mechanical sweeper.
Johnston Engineering Ltd

Above: **An Elgin Crosswind air regenerative sweeper.** *Johnston Engineering Ltd*

Below: **California Transport (Caltrans) V4.**
Johnston Engineering Ltd

Above: **An early Johnston Engineering Ltd unit from 1915.** *Johnston Engineering Ltd*

Above: **Elgin three-wheel mechanical sweeper.** *Johnston Engineering Ltd*

Above: **An Elgin Whirlwind mechanical sweeper.** *Johnston Engineering Ltd*

Above: **A Shelvoke Revopak on duty for Armagh District Council. The 16-ton gvw (gross vehicle weight) Series 18 refuse collector comes with a tilt crew cab.** *Colin Humble*

Left:
This Shelvoke Maxipak utilised the PYC Series crew cab.
Colin Humble

Below:
No less than nine Albion Clydesdale water tankers were put into service by the City of Sydney, Australia. Each was equipped with 1,500-gallon tanks and auxiliary pumping equipment by G. H. Olding & Sons Pty Ltd of Sydney and all are used to keep the city's streets clean.
Ian Allan Library

Above: This 4,000-gallon vacuum cesspit emptier was mounted on a frameless semi-trailer and powered by a Seddon 16:four 28-ton gross tractor unit. The tank section was built by Eagle Engineering Co Ltd of Warwick. It was being used by Chichester Rural District Council in August 1968. *Ian Allan Library*

Above: The Roads & Bridges Department of Cheshire County Council added three Ford D500 gully emptiers to its list of Ford vehicles. They were based at their depots in Preston Brook, Northwich and Upton-by-Chester. The order of 10 additional 'D' series vehicles was part of a fleet of over 80 Fords in operation by the Council. *Ian Allan Library*

Left:
A Johnston road sweeper mounted on a Ford chassis at work in and around Manchester on 6 July 1971 for Manchester Corporation City Engineer's and Surveyor's Department.
Mike Pilkington, Manchester City Council

Below: **Manchester Corporation's Cleansing Department was holding its Annual Vehicle Parade on 18 September 1971. Refuse collection vehicles of all shapes and sizes took part, along with tippers and other vehicles used in the day-to-day management of household waste collection, street cleaning and gully emptying.**
Mike Pilkington, Manchester City Council

Right:
The Cargo range from Ford was represented by a 13.2-ton 1313 with Lacre road sweeper bodywork and a Johnston Sewer Jetter on a 1311 chassis at the Commercial Vehicle section of the Public Works Exhibition in 1982. *Ian Allan Library*

Left:
The Ford D1614 chassis cab powered by the Ford 6.0-litre Turbo II diesel engine is fitted with a Vactor 810 Jet Rodder sewer cleaning device. *Ian Allan Library*

Right:
An Austin lorry equipped with mechanical street cleaning equipment was being used by the Borough of Radcliffe Cleansing Department in January 1955. *Ian Allan Library*

Above: This cesspit emptier was built by N. V. Geesink of Weesp in Holland for work in the Haarlem Municipality. The Bedford 179in-wheelbase normal control 7-ton chassis was powered by a Bedford 330cu in diesel engine. *Ian Allan Library*

These 1960 pictures show clearly how a waste container bin is hoisted and emptied into this refuse collection vehicle. These methods are very similar to those used by the Pagefield systems of the 1940s and 1950s on the Paladins. *Ian Allan Library*

Above: Dowsett, a well-respected name in civil engineering for many years, was the operator of this Ford Lacre compact roadsweeper. Built onto a Ford D1210 108in-wheelbase chassis and featuring Lacre's Compact body, it offered dual-drive controls (left-hand drive as well as right-hand). The photograph was probably taken during the mid-1970s. *Ian Allan Library*

Above: This Bedford lorry in the KM tipper range, seen in May 1967, was one of the first Telehoist Load Lugger bodies on the KMS chassis. It is being operated here by scrap merchants from Witney in Oxfordshire on skip duties. Alongside is a Jones KL44 mobile crane. *Ian Allan Library*

Left: This Dempster DB30 Dumpmaster Compaction body is mounted on an AEC Mammoth Major eight-wheel 24-ton gvw (gross vehicle weight) chassis. The 30cu yd nominal capacity body can handle up to 75cu yd of loose materials by compaction with its 25-ton hydraulic ram. Discharging is done by releasing the rear doors and ejecting the material out of the vehicle's body. It is seen here working for the Borough of Shoreditch, Cleansing & Transport Department, in December 1962. *Ian Allan Library*

Right: Garwood has been producing winches, cranes, logging equipment, bulldozers, scrapers and a whole host of other implements for many years. In this photograph Argyll and Bute District Council is operating one of the two Ford/Garwood FL3000 front-loading refuse collectors. This was the first time that this type of body had been specified by a local authority and fitted on a Ford chassis. The D2418 6 x 4 chassis, powered by the Cummins 8.3-litre V8 diesel engines, could cover up to 1,000 miles a week emptying large refuse bins sited at hotels, caravan parks, lay-bys, ferry terminals etc. The relatively light 24-ton gvw Ford chassis takes 30cu yd of refuse compacted to 500/700lb/cu yd with a payload capacity to spare and the powerful 170bhp Cummins engine, coupled to a Ford six-speed all-synchromesh gearbox and two-speed rear axle, gives the loaded vehicle an excellent range of performance. The complete cycle of picking up the bin and emptying it is controlled by the driver whilst still seated in the cab. A special 'periscope' mirror mounted in the cab roof allows him to closely control the discharge into the hopper. The load is ejected by a hydraulic ram through the top-hinged rear door in the normal way. Ford truck specialist dealer James A. Laidlaw Ltd of Airdrie supplied the chassis and Scapa Engineering Ltd of Blackburn fitted the body. *Ian Allan Library*

Right:
A 1924 Atkinson vertical-boiler under-type steam engine was used as the power unit for this gully emptier, which developed 70hp. The tank and body were also built by Atkinson's. The gross vehicle weight was about 10 tons.
Mr Richard D. Grey, Seddon Atkinson archives

Left:
The States of Guernsey local authority chose a Seddon Mark 5 gully emptier in 1952. The vehicle had the familiar Seddon cab and was powered by a Perkins P6 engine.
Seddon Atkinson archives

Right:
This Mark 758 Seddon 3-ton-capacity side-loading refuse collection vehicle, powered by a Perkins P4 engine, was 5 tons 10cwt gvw. The cab and body was designed and built by Seddon. This particular vehicle was about to be exported to Hamilton, Bermuda in 1953.
Seddon Atkinson archives

Above: Almost all of the street lighting maintenance was the responsibility of the local authority until contracts were issued to the electricity suppliers. This Seddon Mark 758 5½-ton gvw vehicle (including 3-ton load capacity) was supplied on 5 May 1953 to Glasgow Corporation's Lighting Department. Powered by a Perkins P4 engine it was equipped with a manually hand-cranked extending tower to give access to street lights of varying heights. *Seddon Atkinson archives*

Left: A Mark 7 Seddon with a Compressmore body being used in the mid to late 1950s by Blackpool Corporation's Cleansing Department. *Seddon Atkinson archives*

Right:
Seddon produced the Mark 15 in 1958. It was powered by a Perkins P6 engine and had a gvw of 12 tons. This vehicle was being used by the Borough of Lewisham for the street collection of refuse. The Compressmore 26cu yd body, built on a Seddon chassis, included a 6/7-man crew cab with a high roof and an inbuilt wet clothes locker. *Richard D. Grey, Seddon Atkinson archives*

Left:
A Seddon Mark 758 from 1958 is being used here by the Urban District Council of Cheshunt on street sweeping duties. The vehicle features a Seddon cab and chassis and a Brockhouse Sweepmaster body. *Seddon Atkinson archives*

Right:
This Seddon 13 equipped with a Seddon Shark refuse body had a gvw of 15 tons and was powered by a Perkins P6-354 (120hp) engine. It is being used by a contractor in the Clapham District of London. The Seddon Shark would normally be equipped with a full crew cab. However, when container-lifting equipment was fitted, only two crew were needed. *Richard D. Grey, Seddon Atkinson archives*

Above: This Seddon 13/Four gritting lorry was powered by a Perkins 6-354 engine. The vehicle had a gvw of 15 tons and was equipped with a road gritting body built by Atkinsons of Clitheroe, Lancashire (no relationship to Atkinson Lorries (1933) Ltd, with whom Seddon joined forces in 1970). *Seddon Atkinson archives*

Above: The Mark 2 version of the original Seddon 13/Four municipal vehicle, seen here around 1976. The cab was supplied by Motor Panels without the roof which was supplied by Seddon, whose high roof allowed the six-man crew to walk in. The body was supplied by Jack Allen. The grille logo on the front — SAM — stands for Seddon Allen Municipal, not Seddon Atkinson Municipal as is thought by many in the industry. *Seddon Atkinson archives*

Right:
Launched in 1978, the Seddon Allen Municipal (SAM) 200 series was powered by the International Harvester D358 diesel engine of 107ghp (gross horsepower). Fitted with a Jack Allen body, it had a gvw of 16 tons. In 1974 Seddon Atkinson was acquired by International Harvester, just four years after Seddon had acquired Atkinson. *Seddon Atkinson archives*

Left:
This Seddon Atkinson 200 Series municipal vehicle, powered by a D358 engine from International Harvester, is fitted with a Glover body. It is seen here working for West Somerset District Council in 1980. *Seddon Atkinson archives*

Above: This Seddon Atkinson 201 Series municipal vehicle had reverted to a Perkins P6 354.4 engine of 112ghp or an optional Perkins T6 354.4 giving 143hp. The gvw of the vehicle was 16 tons. It is seen here while working for the Argyll and Bute District Council in 1982. *Seddon Atkinson archives*

Below: Seddon Atkinson and Jack Allen supplied this vehicle in 1984. The 201 Series 6 x 4 was powered by a Perkins T6 354.4 engine of 148.6bhp, equipped with a Jack Allen Collectomatic body. *Seddon Atkinson archives*

Above:
The interior of the cab on a Seddon Atkinson municipal vehicle. It has facilities for four crew members, including a driver and passenger seat in the front. There is 90° opening on all doors to allow for easy access and it is equipped with wide steps and a high roof so the crew can stand up inside.
Seddon Atkinson archives

Right:
Jack Allen is one of the foremost suppliers of municipal vehicle bodies. In 1976 it was using this service vehicle, thought to be based on a Seddon Pennine 4 coach chassis.
Seddon Atkinson archives

Above: A Seddon Atkinson 2-11 6 x 4 municipal vehicle on demonstration in 1992. Powered by a Perkins Phaser 180 engine of 173bhp, it featured a three-man day cab. Note that there is no high cab here. It was introduced to coincide with the then-new wheelie bins. A smaller crew was required for these operations. *Seddon Atkinson archives*

Above: The Seddon Atkinson 2-11 is equipped here with a gully emptier for Woodspring District Council, Weston-super-Mare. *Seddon Atkinson archives*

Above: A turbo Perkins Phaser 160T engine is the power behind this Seddon Atkinson 2-11 4 x 2 fitted with a Norlea rear-feed refuse body, working in Manchester. *Seddon Atkinson archives*

Above: This Seddon Atkinson 401 8 x 4 from November 1985 is powered by a Rolls-Royce Eagle 265-litre engine, developing 254 hp. It is fitted with a Jack Allen Collectomatic body for the collection of commercial waste, using a rear-mounted skip loader. It also sports the 'SAM' badge on its grill. *Seddon Atkinson archives*

Above: The Seddon Atkinson SAM model 25-21 of 25 tons, powered by a Perkins Phaser 210 engine, features a mid-non-steered axle. This Jack Allen Collectomatic is here on demonstration. *Seddon Atkinson archives*

Above: The Europacer 300 was based on the wider Strato cab. The larger cab with high roof and wider doorstep access was designed exclusively for refuse collection vehicles. This Seddon Atkinson machine is working in 1988 for the Blackburn-based Noblet, one of the most respected names in municipal services.
Richard D. Grey, Seddon Atkinson archives

Above: A studio photograph of the Seddon Atkinson Euromover, launched in 1999. The cab sits on a lower frame extending in front of the engine, giving a lower than usual step height. Although it is likely to see service in other areas, undoubtedly it will be used by municipal bodies in many instances. *Seddon Atkinson archives*

Above: Code-named 'Oscar' (one-step-cab-access), the Seddon Atkinson Pusher 2000 Incomol refuse collection body is built on the very latest 26-tonne 6 x 2 Euromover chassis. *Seddon Atkinson archives*

Above: A Seddon Atkinson Stato 8 x 4, powered by a Cummins M-113-40 (340bhp) diesel engine. It has a commercial waste skip loader body and is being operated here by Lancashire Waste Services (LWS) working in Preston, Chorley, Blackburn, Leyland and elsewhere in the county. *Seddon Atkinson archives*

Above: A Norba rear-loading commercial and domestic waste body being carried on a Scania 4 x 4 chassis. *Ian Allan Library*

Above:
The Cleanaway Group chose a Heil refuse body complete with forward-loading bin handler/compactor on an ERF EC11 34MU4 to pick up industrial and commercial waste in the Crewe area of Cheshire. The chassis is an 8 x 4 configuration. *ERF Ltd*

Left:
An ERF EM6, with a Jack Allen municipal refuse collection/compactor body, on demonstration. *ERF Ltd*

Left: An ERF EM8 demonstration unit, with a Jack Allen body. The EM6 in the previous picture is a 6 x 6 while the EM8 is an 8 x 8. *ERF Ltd*

Above: This ERF EM8 also has a Jack Allen body. This time, however, the machine is in operation for UK Waste Limited. *ERF Ltd*

Above: A Schmidt SK100 sweeper at Blackpool Pleasure Beach, helping to keep a 42-acre park clean and tidy for over 7.1 million people each year. *Public Relations Department, Blackpool Pleasure Beach*

Above: Kirkless Waste Services operates a range of waste disposal vehicles, including the Volvo 320 (left) and the Foden 3340 (right) seen outside Huddersfield Town Football Club's new stadium. *LWS*

Above: Lancashire Waste Services' Leyland DAF roll-on/off waste container on its 8 x 8 chassis. *LWS*

Left: A small street cleaning vehicle manufactured by Schmidt at work in Lancashire. *LWS*

LWS street sweeping (above) and refuse collection services (left) (operated in conjunction with Cotswold District Council) use these typical modern-day vehicles. *LWS*

Broom can be angled 45 degrees either right or left. Note the view of hydraulic control system for broom angling and raising or lowering of either end of the boom.

Angle and rotation of the broom can be instantly changed from drivers seat—eliminates the need for turn-arounds. Broom as shown is angled 45 degrees to left.

Broom swings on a large, strong turntable welded securely to the frame. The power hydraulic cylinder for angle adjustment moves it 135 degrees and positions it parallel to the frame for transporting.

FORWARD or BACKWARD
without turning machine around

FASTER • EASIER • MORE ECONOMICAL

POWER HYDRAULIC CONTROLLED FROM OPERATORS POSITION

Engineering COMPANY, INC. • 3045 Highway 13. St. Paul, Minn. 55111

Above: In America, things do look a little different, such as this mechanical broom, mounted on a chassis much more similar to that of a motor grader. It is manufactured by LULL of St Paul, Minnesota. *LULL*

Above: Mounted on the front of this US highway truck, this massive mechanically-operated broom built by LULL is intended for cleaning the roads of dirt and even snow. *LULL*

Faun — A Company History

One notable manufacturer of municipal appliances had a somewhat unusual entrance into the industry. Faun was founded in 1940, as a subsidiary of a company called Saunders Roe, which was based on the Isle of Wight. It was founded to repair and refurbish flying boats, mainly the famous Catalinas. Anglesey was chosen, however, because it was less accessible to enemy aircraft and had a calm anchorage on the Menai Strait.

From 1945, after the war, the company switched to commercial and military engineering products, based around its expertise in aluminium alloy welding. At various times it made buses, patrol boats, mobile bridging to cross rivers, aircraft-loading bridges, aluminium roadways (trackway), divers' decompression chambers, and many other products.

In 1969 the company entered the waste disposal industry, signing a licence to manufacture the Shark (later Rotopress) rotary-type refuse collector with Kuka (later Faun when Faun took over Kuka). In 1990 a further licence was signed to mount and distribute the Grange 'Jumbo' packer plate-type refuse collector. It was decided to market it as Variopress, the brand name of the Faun packer plate machine which was made in Germany.

In 1994 the German manufacturing base acquired the licence from the USA to manufacture and distribute the Garwood (later Frontpress) front-end-loader-type refuse collector. By this time Faun was in the receivership of Aspinall, the then distributor for the company. Laird (Anglesey) Ltd (now FMV) took on the licence for the manufacture of the Garwood/Frontpress and over a period of time has reverse-engineered (fitted the rear loading vehicle with a hydraulically activated pressure plate) the product until now only the basic fabrication is still made by the Aspinall family.

In November 1995 Faun acquired Laird (Anglesey) Ltd from the Laird Group plc and the name was changed shortly thereafter to Faun Municipal Vehicles Ltd. Manufacture of the Rotopress ceased in 1996 and mounting and distribution of the all-new Rotopress 2000 commenced. The Variopress 2000 is manufactured at the German company's plant with its state-of-the-art robotic welding technology and the company also has a licence to manufacture municipal vehicles at the company's factory in France, where the company is known as Grange. By 1997 a new model packer plate, the Variopress 2000, was launched on the UK market to sell alongside the French-made Variopress. A new custom-built production facility was opened in January 1998 on the Bryn Cefni Industrial Estate in Llangefni, Anglesey, which mostly carries out final assembly, although some manufacturing of parts and vehicles is also carried out.

Above: **One of the many boats once made by Faun in the early years of the company.** *Faun MV Ltd*

Above: **A Kuka refuse collection body from 1935 mounted on a Mercedes Benz chassis. Kuka later became integrated into the Faun Group.** *Faun MV Ltd*

Above: **In this photograph the City of Glasgow Corporation was using a Laird body, mounted on a Shelvoke & Drewry chassis, for refuse collection.** *Faun Municipal Vehicles Ltd*

Above: This Laird (Anglesey) Ltd refuse collection vehicle was being operated in 1973 back home in Wales. *Faun MV Ltd*

Above: Another Laird (Anglesey) Ltd body, this time mounted on a Commer chassis. *Faun MV Ltd*

Above:
A Laird (Anglesey) Rotopress, mounted on a Leyland Constructor chassis. *Faun MV Ltd*

Left:
A trio of Faun refuse collection vehicles, two of which are built onto Seddon chassis and the third (far right) onto a Leyland chassis. *Faun MV Ltd*

Left:
This time a Faun refuse collection body is being carried on an AEC Mandator chassis for Manchester Corporation Cleansing Department. *Faun MV Ltd*

Above: This Kuka detachable municipal refuse body incorporated a useful device so that it could be left at a central point for loading and be collected for emptying later. *Faun MV Ltd*

Above: Mid-Sussex District Council was the user of this Faun refuse collection vehicle, mounted on a Seddon chassis. *Faun* MV Ltd

Above:
A similar Seddon/Faun refuse collection vehicle to that in the previous photograph, albeit in a somewhat different livery. *Faun MV Ltd*

Above and right:
Two views of a Laird Variopress refuse collection body, this time mounted on a Renault chassis.
Faun MV Ltd

Above:
The ultra-modern Rotopress 2000 by Faun, mounted on a Dennis chassis, is operated here by Wealden District Council. *Faun MV Ltd*

Above and right:
A Scania 93M-280 chassis provides the pulling power for this Laird/Garwood municipal commercial waste collection vehicle on demonstration from Laird Anglesey Ltd. Note the Garwood forward loader, which can lift heavy duty bins from ground level, empty them into the body of the vehicle and return them to the ground in record time. *Faun MV Ltd*

A Laird Rotopress on demonstration. *Faun MV Ltd*

THE ENVIRONMENTAL IMPACT

The environmental impact of waste disposal has become an increasingly important issue. The increasing environmental and actual cost of land use, as well as tipping taxes, have led to a much greater emphasis on the recycling of household and, when possible, commercial waste. Many councils now contract out the collection of waste paper, tins, glass and plastics, with these materials being sorted during the collection process and delivered to premises designed to accept them, before delivering in bulk to the remanufacturing plants. Builders' waste, eg bricks, concrete and asphalt, are now commonly being recycled for use as hardcore on building sites, further reducing the need for vast tracts of land to be used for the disposal of everything — a situation which was once so readily accepted. Many once-difficult-to-dispose-of items are now being accepted for reuse. Substances such as industrial oils and cooking oils can both be used in oil-burning heating systems.

No longer are the massive landfill sites just a convenient place to dump anything and everything. Those that are still very active are very carefully engineered to meet high environmental standards, so that methane and waste fluids (many of which are highly toxic) can now be burned off to provide fuel for electricity generators, rather than flow off into nearby streams or rivers. Indeed, where once it was looked down upon (wrongly) to admit that one worked on refuse collection, or on a 'rubbish tip', now only skilled, trained personnel are able to work in the industry because of the highly technical nature of the vehicles and systems now being used at both the collection and disposal sites. Today's multi-million pound waste industry, much of which is now operated by large nationwide companies, operates within very strict guidelines laid down by statute and the fines for breaches of the rules are very heavy indeed.

Only recently I was talking to someone who works on a large landfill site in the Greater Manchester/Lancashire area. He told me what it is like to have a very large skip full of baked beans emptied in front of you, or around 30 tons of kegs containing the ingredients of curry. These and hundreds of old beds, mattresses, chairs and settees are just some of the items deposited on a refuse tip daily. Whey and other unusable dairy products, vegetable matter and many food items which have gone past their sell-by date are also likely to end up in a landfill site.

Many of the country's largest landfill sites have similarities with open-cast coal mines. They take up, in most cases, many acres of land. Without very careful management they can have a devastating effect on any nearby urban areas, the major problems being water pollution, methane gas build-up and dust, not to mention the inevitable thousands of seagulls and rats. Nobody would think of tipping thousands of tons of commercial and household waste on prime agricultural land but choosing the right kind of areas to use can be difficult. For example, would you consider dumping on a former industrial site (say an old brickworks, foundry, mine site or an old disused quarry)? All would seem ideal at first glance, but what if the land is already contaminated with industrial waste containing toxic substances which might seep out into the local streams or water courses if disturbed? If so, no amount of careful management of the tip would prevent a disaster. Old mine and quarry sites may seem a good idea on the face of it, but what if the geology of the area is such that major fissures in the surrounding rock provide ideal channels through which toxic liquids could escape, to be caught up in the streams and rivers once again?

Another major problem is access to the site by several hundred lorries a week, requiring sizeable roads to be built to accommodate them. Are they expected to pass through villages, towns or built-up areas en route to the tips? That would not be environmentally friendly. So, what is the answer? Certainly a major campaign to encourage the recycling of most of the commercial and domestic waste is a must. Stricter planning guidelines and licensing have already made it difficult for rubbish to be dumped anywhere (known as fly tipping), but it is important that the right measures are taken to identify the most environmentally suitable areas on which to allow tipping. Concerns need to be addressed about the local water quality, dust suppression and the number of lorries (could the traffic be alleviated by making available a suitable rail link, perhaps from major centres?). A well-thought-out plan for the timescale for tipping to be allowed in any one place (five years, 15 years, 30 years or more?) is necessary.

As areas become full they are covered with layers of top-soil on which landscaping can begin. However, the enormity of the many landfill sites dotted around the country can only serve to reinforce the need for a harder look at the recycling of much of our waste. Options such as the incineration of the remainder may help to reduce the overwhelming burden, but this has other environmental consequences.

Above: With the increasing environmental cost of land use, including new tipping taxes, more and more emphasis is being put on the recycling of household and, when possible, commercial waste. Many Councils now contract out the collection of waste paper, tin, glass and plastic, with the materials being sorted during the collection process and delivered to premises designed to accept them, before delivering in bulk to the re-manufacturing plants. The vehicle shown here is a Seddon Atkinson Pacer 240, a 6 x 4 vehicle powered by a Cummins B Series engine, developing 235hp. The City of Chester, which operates it, is one of many authorities which encourage recycling as a means of reducing the amount of material which has to go into landfill sites. *Seddon Atkinson archives*

Above: This photograph from the air is of Whinney Hill landfill site in Accrington, Lancashire. It is owned and operated by LWS (Lancashire Waste Services). *Image Aviation/LWS*

Left:
Lancashire Waste Services (LWS) is removing industrial waste from an aircraft factory near Preston in Lancashire, a rear-loading compactor refuse body mounted on a Volvo 8 x 8 chassis being used on this occasion. *LWS*

Right:
A removable container mounted on an ERF 8 x 8 chassis deals with baked beans and other waste products from this famous factory near Wigan. *LWS*

Left:
A Leyland DAF fitted with a roll-on/off container carrier, enabling the operator to leave the large waste container at a refuse collection point (industrial, commercial or domestic unwanted items can be deposited into the large receptacle). It is then picked up at a later date when full for transfer to the landfill or incineration plant. *LWS*

Above: The extent of landfill sites and the continual work to provide the space needed for thousands of tons of household and commercial waste can be seen in this photograph. The whole of the cell will be lined with clay and terram-type rubber mats which reduce the possibility of water being taken down into the strata, whence it could find its way into water courses and subsequently drinking water. Groundwater from such sites is normally piped away into lagoons where it is treated to remove any toxins. *Mark Harvey, J. McArdle Contracts Ltd/David Dore*

Below: The batters around these huge landfill cells have to be built up and compacted to a very high standard to prevent leakage. *Mark Harvey, J. McArdle Contracts Ltd*

Above: A view of earthworks being undertaken at Risley by J. McArdle Contracts Ltd, one of the leading contractors in waste tip creation and extensions. *Mark Harvey, J. McArdle Contracts Ltd*

Above: In this view of work at Risley, we see special attention being paid to fine-grading to prespecified levels. A rubbish tip it may be, but as with a new road or housing estate, only high quality workmanship is acceptable, unlike perhaps some of the tips in earlier days. *Mark Harvey, J. McArdle Contracts Ltd*

Above: A view of Risley IV cell which will soon be ready to receive its first loads of waste. In around 10-15 years time at current levels this cell will be full and other sites will be required for future use. *Mark Harvey, J. McArdle Contracts Ltd*